I Believe...

The Gift of Certainty

Dean C. Lohse, M.D.

Cover design by Five Js Design, fivejsdesign.com

ISBN: 978-1-7352559-3-4

I have recreated events, locales, and conversations from my memories. To maintain their anonymity, in some instances I have changed the names of individuals and places. I may have changed some identifying characteristics and details such as physical properties, occupations, and places of residence.

Created with Vellum

This book is dedicated to the parents, the teachers, and the pastors who communicate God's love in the person of Jesus Christ through the power of the Holy Spirit to the children of the world.

"It ain't what you don't know that gets you into trouble.
It's what you know for sure that just ain't so."

—attributed to Mark Twain

Introduction

Is there a God, and if so, what is He like? Do I exist as a random biological event? Does my life have a meaning? What does love have to do with anything? Is there really a life after death? Most of us experience uncertainty over the big questions, and the sheer volume of available information doesn't help. Sorting through the noise is difficult. Sometimes the information is wrong, sometimes it is meant for another person or situation, and sometimes people lie—for fun, profit, political power, or influence, take your pick. Others are incredibly ill-informed about everything except their own ignorance. Yet others suffer from some form of mental illness that includes an urge to share a bizarre viewpoint with the world. We are subjected to increasingly loud and insistent opinions that are usually wrong. We are forced to find our own truth.

I am certain about some things. My journey has taken several decades that have included hopes and dreams, loves lost and loves held dearly, dismal failures and spectacular successes, a long career in the difficult healing discipline of neurosurgery, some minor skills as a builder of small boats, plus many days and nights spent in the wilderness. I've been influenced by my education and experience in science, what I've learned about philosophy and mythology, and some biblical wisdom. The result is what Mark Twain allegedly referred to as "what you know for sure."

Unlike Mr. Twain, I expect that my things that I "know for sure" will not be what gets me in trouble. In the past, I've listened to the noise and believed other things that just weren't so. To keep me on track with the truth, I have framed my journey by the ancient and enduring wisdom of the Apostles' Creed. The result is the gift of certainty.

I hope "I Believe" will inspire your journey.

The Apostles' Creed

I believe in God, the Father almighty,
creator of heaven and earth.

I believe in Jesus Christ, his only Son, our Lord,
who was conceived by the Holy Spirit
and born of the virgin Mary.
He suffered under Pontius Pilate,
was crucified, died, and was buried;
he descended into hell.
The third day he rose again from the dead.
He ascended to heaven
and is seated at the right hand of God the Father almighty.
From there he will come to judge the living and the dead.

I believe in the Holy Spirit,
the holy catholic[1] church,
the communion of saints,
the forgiveness of sins,
the resurrection of the body,
and the life everlasting. Amen.

1. Or "universal."

Part One

I believe in God,
the Father almighty,
creator of heaven and earth.

Chapter One

"I believe in God"

My father's father never went to church. I did not know him very well. He died when I was nine and his last years were marked by hearing loss, which made it hard for him to understand, and Parkinson's disease, which made it hard for him to be understood. But by all reports he was a good man: hard-working, honest, loyal to family and friends, a good husband, a good father, a good farmer. I don't know exactly what he believed; probably nobody did. He never complained and he never explained. He lived as if he believed that a small corner of a cold, unfeeling world could be brought to heel with diligence and luck, and a better world would be left behind for his wife and children.

Yet, on his deathbed after his wife and children had gathered and prayed, he had a few moments alone with his son, my father, and he said, "Do you think there could be something to all that?"

It's a good question, better asked earlier in a lifetime

because it might make a difference not only in how you spend eternity but in how you live on Earth. Yet, belief in God is often difficult for highly practical, self-reliant men like my grandfather. So, men like him don't talk about it. They let others (like my grandmother) worship and pray while they go about their business, possibly entertaining but never answering that deathbed question. *Could there be something to all that?*

I wonder why he waited until this moment to ask that all important question. He had lived a hard life, as had the generations of men before him. The survival of his family depended on his attention to the details of the here and now, and I doubt he allowed himself the luxury of wonder about the what-ifs and the world to come.

Yet, he must have wondered at the stark beauty of a winter dawn, the slow pastel beauty of a Minnesota summer sunset, the magic of new life at the birth of his children, the mysterious feeling that death cannot possibly be the end when his own parents died. He must have wondered where this fraught, lovely world had come from, and if he had a destiny and a purpose.

He was a decent man, and he may have wondered where this universal sense of decency came from. We all know right from wrong, and we all fail to get it right every time because we all experience the tension between doing what we know is right and what we think is likely to be good for us. Decency is counterintuitive to self-preservation, but there it is, and Grandfather chose decency.

Decency with its attendant courage, and beauty and justice and truth don't have a place in evolutionary theory. There we are the product of "the selfish gene" and our species progresses by "survival of the fittest." Decency and our other core values are not something that can be proven, but we all know them and treat them as self-evident and real. They come from somewhere holy. Every culture known to history or anthropology treasured these same core values and associated them with God. We are at the end of a chain of wisdom tens of thousands of years old that believes in God.

God challenges us to see Him or ignore Him when we witness acts of selfless courage, scenes of stark beauty, or experience the transcendent during the three big life events: falling in love, the birth of your own child, the death of a loved one. And, if we are fortunate, He may challenge us one more time on our deathbed.

Could there be something to all that?

I answer *Yes*.

I first *experienced* God when I was fourteen. I say "experience" because I already had a "knowledge" of God in the Protestant Christian tradition through Sunday School, the *Bible Story Book* read to me by my mother before I myself could read, and confirmation classes that took three hours of every Saturday morning for two years before the ceremony to confirm my baptism vows and take responsibility for my own soul. Lots of knowledge; not much experience.

Knowledge is a tool, something that can grow or be modified by new information. Knowledge is not personal, not foundational. It is used to solve a problem or answer a question, or it paves the way to frame another question, but knowledge only defines a framework in which we interpret reality. Experience is what defines reality.

By this time in my young life I had knowledge of the Christian God, but I also had sufficient knowledge of the world to be skeptical, and it is my nature to be skeptical. The entire Christian tradition could be a myth no different from the myths of the Romans and the Greeks, and for that matter the myth of Santa Claus—namely, constructs created by society to control the behavior of small children and gullible adults. Karl Marx called religion the "opiate of the masses." Sigmund Freud speculated that God was a construct of our unconsciousness that represented our inner father figure, that part of our unconscious that makes us behave. Richard Dawkins considers God a delusion.

Maybe, maybe not, I thought. But, on the other hand, all those who I loved and respected—except for my grandfather—believed in this Christian myth.

Then I found myself walking alone in a cold, October rain on a deserted city street, severely underdressed in a T-shirt and blue jeans, wet through to my underwear. I still had a long way to go before I got home, so I decided to put God to the test. I asked the Creator of the universe to make me warm and dry, and since the request was de novo I thought it only fair that He should get a full ten seconds to fulfill my request. If He could do that, I told Him, I would believe the whole God, Jesus, Resurrection thing as true. If it didn't happen, I would remain skeptical.

I counted down from ten to zero and was about to embrace the life of an agnostic cynic. The words were already formed in my mind: *See? There is no God.*

At that moment, before I could embrace the words as a thought, much less speak them aloud, the family car pulled around the corner, driven by my brother who didn't actually have a driver's license. He popped open the door and yelled at me to get in, which I did, and I found the car had already been warmed to about 90°F with the heater still blasting away. In a few minutes I was home, changing into dry clothes and eating fresh-baked bread at the kitchen table.

It seems like a trivial incident, a coincidence perhaps. But the unlikely circumstance of my "salvation" occurring at the exact moment I requested it puts the experience beyond coincidental into miraculous. A small and very personal miracle perhaps, but still a miracle.

To say that this experience made me a good person would be a lie on so many levels. This experience did convince me that whatever else I might be cynical about and whatever decisions I made or actions I took, I could no longer tell myself that God was not real. Looking back on this watershed experience, my personal testimony, I think my trivial request may have made God laugh.

But now when I think about this story—a kid alone, cold, wet, and a long way from home—I see it as a metaphor. That's where God finds us, or more accurately since He hasn't lost track of any of us, where we are open to finding God. Like my grandfather—in pain, certain he would not be going home, and uncertain of what lay ahead.

My dad told my grandfather, "Yes, there is something to all that." And my grandfather listened, and a few hours later he died.

For months afterward, my dad grieved over the loss of his father. He sat with his mother every night for several weeks, then called her every evening for the next six months or more. But he remained troubled, slept poorly, and grieved until one night he saw his father in a dream. Grandfather didn't say anything that Dad could repeat later, he only indicated that everything was all right. He was at peace and my dad could be at peace too.

So, "I believe in God," for lots of reasons, but mostly because He shows Himself to lonely, cold, wet boys a long way from home and to old men on their deathbeds. If you seriously ask, "Is there something to all that?" He will show Himself to you.

Chapter Two

"The Father Almighty"

My dad jingled. He took the bus downtown to work each morning and came home predictably at the same time each afternoon to the same bus stop. The stop was three blocks away, so too far to meet him there, but my brother and I were allowed to watch from the front sidewalk. When he came into view on the other end of the block, about two hundred yards away, we could run to meet him. This probably served my mother well; she could have us out of the house, our attention fixed on something not likely to cause harm to ourselves or others, and reasonably safe.

Sometimes on a good day he would let us take turns and ride on his shoulders the rest of the way home. But since it was an uphill climb and there were two of us, usually he would convince us to walk by his side and hold his hand. The top of my head came to about his waist, so my ear was right next to his pocket. What I remember most about those walks was the jingling of change and keys in his pockets. It sounded like little silver bells. Only dads jingled. Not moms, not kids—only dads. My dad.

The Apostles' Creed identifies God as "the Father." God is the Father, but so is the priest, so is the rote beginning of the Lord's Prayer, and so is my dad. So, what can "God, the Father" possibly mean?

In religions other than Christianity, and even in our own Old Testament, God's title is some variation on "Ruler of the Universe" or "Lord and Master." The Man Upstairs, the Big Boss. Not Dad.

"Dad" implies something much different from "Ruler of the Universe," does it not? The Ruler is almighty certainly, but He rules over the nameless masses with the threat of punishment or even complete annihilation for disobedience; the less well-defined benefit of obedience is continued existence. This is understandable. A Being as powerful and universal as we imagine God would not, to our human way of thinking, be capable of being personal. I mean, He's got over 6 billion living souls on this planet alone, not to mention those already in heaven and hell. And we are not even talking about managing the couple hundred billion galaxies in the observable universe and all the living creatures great, small, and microscopic. It's a helluva job and no reason to think He would pay any attention to me (unless maybe I stepped out of line).

Dad, on the other hand, knows exactly who I am. He created me with his own DNA through his passion and his love. He knew me as a baby before I could know myself. He protected me and guided me until I could take care of myself. And every now and then he punished me for being violent or disrespectful or putting myself or others in danger.

Jesus shocked the world when He referred to God as

"*Abba*,"[1] an informal and familiar Aramaic term meaning something close to "Dad." God is personal; He is loving. After understanding that He really exists, the most important thing we need to know about Him is this: He's Dad. He created you with His own DNA through His passion and His love. He knew you before you knew yourself. He protects you and guides you. If you are becoming a danger to yourself or others, He will punish you.

And, yeah, sure, He's almighty. Not some little god like in Greek mythology that has control only over a little arena. He's the big God, the only real God. We kinda expect almighty out of Him; otherwise, He wouldn't be God. But when we put *Father* together with *Almighty*, I think the most amazing thing about all the powers He displays is to be Dad. Not just to me, but to every human being born since the beginning of time. The Father Almighty. Dad.

Now I'm an old man, and I've lived a life that was far less than perfect. I don't want to label myself a sinful man, but I can't claim to be righteous either. When I've drifted, I've struggled and felt unsettled, lonely, and lost. Sometimes I have been a danger to my own soul and the well-being of others, and for that I have suffered. On other days, my best days, I wait for Dad, looking for him in the distance, and when I see Him coming, I run as fast as I can to meet Him. And we walk together, holding hands, I hear a jingle like silver bells in my ear because only He sounds like that. And together we walk toward home.

---

1. Mark 14:36.

Chapter Three

"Creator of heaven and earth"

Held in secret for over half a century, here is the confession of a true nerd: When I read the genetics section of the Biology 101 textbook and came to the part about the discovery of the DNA molecular structure, I cried tears of joy. It was so beautiful, so intricate, so unexpected. So perfect.

I know, it seems like a nutty response. Me, sitting in a corner of the Wilson Library at the University of Minnesota, sniffling over a textbook. That's why I never mentioned it until now. But today maybe some of us should remember that the world around us is all those things—beautiful, intricate, unexpected, so perfect—in a word, miraculous.

And this is important, why?

Because if you understand the world as miraculous, you understand yourself as a miracle, a creature of destiny and purpose. If you are content to limit your understanding of the universe to the description of the mechanical moving parts and deny the spiritual forces that created those moving parts, you will understand yourself as a random being in a purposeless universe. You may call yourself a scientist of sorts, but you will have given up the greatest purpose of science, which is to understand the miraculous. More important, you have given up on your own destiny.

Our best science has always come from people seeking the truth about creation and, therefore, the truth about God. Copernicus, a Catholic scholar in the sixteenth century, is the scientist who demonstrated that the Earth revolved around the sun, leading to our future understanding of our solar system, our place in the galaxy, and our galaxy's place in the universe. Georges Lemaitre, another Catholic scholar working in the twentieth century, was the first to formulate the theory he called "The Beginning of the World," and what we now commonly call the Big Bang theory. (Edwin Hubble confirmed his calculations two years later.) And in modern times we can look to Frances Collins, MD, PhD, who led the Human Genome Project and was director of the National Institutes of Health from 2009 to 2021.[1]

The Bible begins with the first miracle from which all other miracles follow. The first verse, Genesis 1:1, tells us that "In the beginning God created the heavens and the earth."

The phrase is so familiar that it is easy to skim over. But the verse says two important things that are not givens but are confirmed by twentieth-century science. The first is that there is a beginning—a beginning to time and a beginning to space and a beginning to matter. Alterna-

1. For a full review of his faith journey, see Francis Collins, *The Language of God: A Scientist Presents Evidence for Belief* (Free Press, 2007).

tively, some religions propose that time is circular, that we are on a "wheel of time." Other religions skip a true origin of time and space, apparently assuming that these things have always been around, and then start a myth with the creatures or gods that populate the earth. Genesis clearly says there is a beginning.

The second thing this verse says is: "God created."

Compare this first verse of the Bible with what creation science tells us. The Big Bang theory proposes that time and space and matter indeed had a very specific beginning, and a power beyond our comprehension brought an ordered universe into being out of nothing. Science tells us about the *how* but never about the *why*. Science describes what the power beyond our comprehension did but doesn't call that power God. The Bible doesn't have to limit itself to description; it can go on to the why and to the truth.

The first verse of Genesis doesn't give a timeline, but our best scientific guess is that the "big bang" occurred 14.5 billion years ago. After the big bang, say about a billion years or so, galaxies like our Milky Way started to form. After another 8.5 billion years, our own sun took shape. Another 3 billion years later, the Earth was formed. That would be about 4.5 billion years ago.

The biblical "first day" had not yet occurred. Genesis 1:2 tells us that "Now the Earth was formless and empty, darkness was over the surface of the deep, and the Spirit of God was hovering over the waters." Creation science describes the primordial Earth less poetically, but only a little imagination is required to envision the same thing. Our atmosphere was primarily water in the form of steam, and the surface was an ocean of liquid magma (molten rock)—that is, darkness over the surface of the deep.

Genesis 1:3 tells us "God said, 'let there be light,' and there was light." Science tells us that the earth cooled, the steam atmosphere condensed to water and clouds, and sunshine reached the surface. In other words, there was light.

Genesis 1:4 tells us "God saw that the light was good, and he separated the light from the darkness." Science tells us that the Earth rotated, perhaps at the same rate that it does today, perhaps not, but clearly in a rotating pattern of light and darkness. This is not true of our moon and not true for all planets; it is true specifically for Earth.

Genesis 1:5 tells us that "God called the light 'day,' and the darkness he called 'night.' And there was evening, and there was morning—the first day." From this point on we can argue what the term *day* might have meant in prehistory. Is a day a prescribed twenty-four-hour period? Was it a single rotation of our planet, and if so, did our planet rotate at the same interval a billion years ago as it does now? Or is a day a metaphor for an era or an eon?

I personally don't feel the need to argue about the definition of *day* in this context, but what is inarguable is the sequence of events described in Genesis 1:1–27. The waters separate from land, vegetation appears, the sun and the moon and the stars become visible, sea creatures populate the oceans, then birds appear, then land animals, and then, finally, man.

And this fact is in itself amazing: Our most ancient record of our creation in the book of Genesis outlines exactly the sequence of events that modern science confirms, the same separation of waters and land, vegetation, then sea creatures, then birds, then land animals and then, at last, *Homo sapiens*, mankind. In other words, creation science confirms what the book of Genesis has already told us.

I like to imagine God the Father conveying a vision of His creation to an individual or group of individuals before they had the mathematics and the telescopes and the language to figure it out for themselves. Because it was important. Because each individual was created in the image of God with purpose and destiny, and they should know that first. God conveyed a vision of His creation that looks like what modern creation science has discovered, or more accurately, rediscovered.

Abiogenesis is the study of the beginning of life on earth. It is not hard to imagine, or even reproduce in the laboratory, primordial elements forming amino acids and amino acids forming simple proteins. But in biological life, proteins are designed by and produced by nucleic acids in the form of DNA and RNA. Mitochondria are necessary for energy production and to transmit their own generational information. Complex protein structures are needed for the simplest of cell walls in even single-cell organisms. These processes are not easy to imagine happening as a matter of chance. The appearance of nucleic acids and cell walls and mitochondria is as likely as putting chunks of aluminum, copper, plastic, and silicon in a bag, zapping it with a bolt of lightning and expecting a MacBook to appear. The beginning of life on Earth was not a matter of chance; it was a miracle, an act of creation.

This miraculous creation in original form is hypothesized to be a single-cell organism with about 355 genes termed the Last Universal Common Ancestor (LUCA). From this organism came all of the microbe world, all of the plant world, and all of the animal world. The process began about 4 billion years ago, presumably through the process of evolution.

Evolutionary theory implies chance. Random environmental factors pressure change, the "selfish gene" surviving to the next generation, the "survival of the fittest," and somehow the fittest become a new species. The evidence for evolution in the fields of comparative anatomy and fossil science is overwhelming. Furthermore, evolution has a certain elegance, an elegance that comes from an understandable explanation of complex observations. The forms of life on our little planet are not only incredibly diverse, but also they are inextricably related.

I don't argue against evolution. I argue against chance.

Every new species, from LUCA to *Homo sapiens*, is not a result of random environmental events but an act of miraculous creation. I say this for three reasons. First, evolutionary theory outlines a system that goes from the simple to the complex. This goes against our general experience that systems tend to disorganization without continuous care and creative leadership. Buckets rust and leak. A clear corporate leadership plan erodes and the vision is distorted. Random events and chance never lead to improvement, higher complexity, and greater creativity. Entropy reigns.

Second, the spontaneous creation of a new species seems to be extremely rare. Our careful observations of the natural world over the past two centuries have documented many cases of extinction of a species—from the smallpox virus to the carrier pigeon to the northern white rhinoceros—but no *new* species. If evolution works, as we assume, to expand biodiversity, why are we not seeing more new species than we are seeing extinctions? In microbiology labs we can grow literally billions of replicating generations of, for example, *Escherichia coli* bacteria; they may have great genetic variation, but they are all *E. coli*. They never turn into *Klebsiella* or *Staphylococcus*. Our genetics labs have grown many generations of *Drosophila* fruit flies, but they are all still fruit flies. Our biomedical labs have grown millions of generations of laboratory rats to explore various genetic questions. The rats have tremendous potential for diversity within their species, but they always turn out to be rats—never mice, never cats. If in all our observations in the laboratory and in agricultural breeding programs we have never observed a new species, should we not assume that the creation of such is so rare that it is not a matter of chance but a matter of the miraculous?

Third, the process of creating each new species by evolution is beset by such necessary rare events that it must be considered miraculous. Evolution *within* a species is a no-brainer. Species adapt genetically to their environment or they become extinct. Our earliest recorded ancestors knew this and bred domesticated animals as part of our own agricultural revolution to begin our human civilizations. We breed dogs as diverse as a Chihuahua and a Great Dane, but they are all still dogs. We breed longhair, shorthair, long-tailed, or short-tailed cats, but they are still cats. Darwin observed many kinds of finches and tortoises, but that was nearly two centuries ago, and they remain the same species of finches and tortoises.

But evolution from one species to a totally new species is much more complex. Imagine a breed of sparrows threatened with extinction by the onset of global warming. Fortunately, a mutation occurs, and one bird has lighter feathers and is more immune to the effects of heat. Great! Now all that must happen is that another bird in the same generation of the same flock has the same mutation, and these two find each other and mate. Then their offspring must mate with each other without bringing forth another trait that is fatal, and so on until there exists a sustainable breeding colony. This is possible, but it is not probable. It is not even likely. It is miraculous.

And because the evidence for evolution is so great, we must conclude that miracle has occurred countless times accounting for each new species over the past 4 billion years.

I have a choice to believe in chance or to believe in miracles, and that choice is the difference between hope and despair. Because it's a big universe out there, and it is easy to believe that I am a random speck on a little planet circling a middling star that is one of 200 billion stars in only one of more than 200 billion galaxies, and that is only in the observable universe, and biological life came to be out of a chance mixture of chemicals and then randomly evolved into humans, of whom I am but one of billions. Yes, I am a speck of protoplasm in a big universe existing at the end of a 15-billion-year time span.

But I am not random chance.

That big, big universe? Created or accidentally formed? What was there before the big bang? What exists beyond the observable universe? What lies beyond the Higgs field through which all matter must pass to exist in this space-time continuum?

To argue against a creative force behind the universe, our Earth, and biologic life would necessitate some twisted logic. We can call this "Creative Force" just that and leave it there, saying, "How could we as mere mortals understand such a magnificent power?" There is honesty in that. The created can never fully understand the Creator.

But we know better because of what we know about ourselves. The Creative Force is God, and He does indeed instill in us, the sentient beings in His creation, a sense of awe and wonder, and yes, purpose. We will never fully understand Him, but in our best moments we will seek Him.

I haven't gotten too emotional about molecular structures lately, but I have been awestruck by quiet, northern lakes at dusk, raging ocean waves on the edge of hurricanes, and the sky on a clear, moonless night in the wilderness. These are mysteries and miracles. They strike me with awe. I am driven to humility. I am conscious that I am made up of billions of delicate and intricately designed molecules. Along with the universe, I am miraculous. I am created.

Part Two

I believe in Jesus Christ,
his only Son, our Lord,
who was conceived by the Holy Spirit and
born of the virgin Mary.
He suffered under Pontius Pilate, was
crucified, died, and was buried;
he descended into hell.
The third day he rose again from the dead.
He ascended to heaven
and is seated at the right hand of God the
Father almighty.
From there he will come to judge the living
and the dead.

Chapter Four

"I believe in Jesus Christ"

Jesus existed and exists.

He existed as a historical figure, showing up outside religious literature in the history of first-century Rome as recorded by both Tacitus in his *Annuls* and Josephus in his *Antiquities of the Jews*. The Synoptic Gospels (Matthew, Mark, and Luke) have remarkable correlation with each other and are consistent with known historical events of the first century AD. Evidence of the historical existence of Jesus is as great or greater than that of any historical figure of His day.

Jesus existed and exists as a personification of our humanity. He is the counterpoint to the "law of the jungle" and the "survival of the fittest," the rules we learn on the playground and in the workplace, and the assumptions behind the theory of evolution. We learn that if we want to be leaders, or even chosen for the team, we need to be self-confident, emotionally secure, and strong. Then we learn to be competitive in school and in the workplace by adjusting our ethics to the situation, demanding accountability, compromising when necessary; winning is everything and the only thing. And peer approval is critical. To master these skills is to achieve worldly success. If we come up short, as most of us do, we find our niche and seek consolation.

Then Jesus comes along with His "Blessed are's." He blows off all the aforementioned skills. He says if we are poor in spirit, we receive the kingdom of God. If we are meek, we will inherit the earth. If we mourn, we will be comforted. If we forget about adjusting our ethics and compromising our principles and simply hunger for righteousness, we will be satisfied. He gets crazier and crazier until He says that if we are insulted, persecuted, and lied about we should be exceedingly glad.[1]

Nobody wins in this world by starting with "poor in spirit" and ending with "persecuted because of righteousness." Yet somehow all our worldly successes are empty and meaningless unless that is exactly where we start and end. Jesus exists as the philosophical and spiritual underpinning not of our worldly success or our survival, but of our humanity.

Jesus existed and exists as the central figure in the best of our stories, our myths. You know the plot of *The Lord of the Rings, Star Wars*, or *The Hunger Games*? It is the ancient and universal story of the hero's journey. It pervades blockbuster movies of the past century as well as the myths and legends of all cultures at all times in recorded history. Jesus' life is the same plotline: He came from humble origins, was chosen for an impossible task, and gathered chosen comrades. Their early successes surprised the unbeatable foe, who gathered the forces of evil toward a climax where all hope seemed lost. But

1. See Matthew 5:3–12.

against all odds, victory and salvation of the world was won by the force of good and the faithfulness of the hero. We all know what happens when Luke Skywalker follows the mysterious command to "Trust the Force." We cheer Frodo Baggins as the Ring, Gollum, and Frodo's finger are cast into the fire at the Crack of Doom. That is the echo of our ancient myths manifested in *movies*. When Jesus says, "Thy will be done,"[2] it is the echo of our ancient myths manifested in *history*. Jesus' life is the historical fulfillment of our universal myth.

Yet with all that, Jesus existed and exists in a way that is even more convincing. He appears personally in lives. Saul met him on the road to Damascus, Augustine met him in the dusty streets of Carthage, and St. Francis met him in a chapel in Assisi. An angry and profane sailor named John Newton met him on a slave ship during a storm in the North Atlantic and went on to write the beloved hymn "Amazing Grace." You know the words: "Amazing grace, how sweet the sound, that saved a wretch like me."[3]

These encounters are memorable but not unique. Jesus appears to people not only in historical accounts but in everyday lives, usually at a time of deep personal distress—a life-threatening situation or illness, addiction, worldly failure or moral turpitude, or intense loneliness. Sometimes, like in Saul's case, He appears before an angry man to turn a life around.

I was about thirty years old, a senior resident in neurosurgery, in the fifth of six years of training at Yale University. My wife and I had been married for nine

---

2. See Matthew 26:39; Mark 14:36; Luke 22:42.
3. John Newton, "Amazing Grace" (1779).

years and had two sons, ages three and five. We had no money. I owed $7,000 ($27,000 in 2023 dollars) in student loans for medical school and my $17,000/year salary barely covered living expenses. My long hours made it mandatory for my wife to endure long hours as a stay-at-home (effectively single) mom if our kids would have anything resembling a normal childhood. I worked eighty to a hundred hours per week, spending every second or third night in the hospital.

The hospital provided a kind of emotional release—I was so busy taking care of sick people I didn't have time to think about myself or my family or our finances. Work was a totally different reality; I would be one kind of person there and a quite different person at home. This sort of emotional detachment is not all bad. In fact, it is necessary to function in either world. But the trade-off was not equal. I could be a humane neurosurgeon because I had a loving family, but I wasn't necessarily a good husband and father because I was a skillful neurosurgeon. Many nights I got home so exhausted that all I could do was eat, sleep, then get up and do it again. Frequently, I fell asleep reading a bedtime story to my children. Other nights I fell asleep at the hospital barely remembering that I had a family.

Though I was on the verge of professional success—in another eighteen months or so I would get a real job with a real salary and a real future—I was also on the verge of personal failure. Because of faults that were entirely my own, our marriage had failed. My relative morality was one of convenience—I did the right thing only when it cost nothing and the wrong thing when it was convenient. I decided. And I believed that what happened in one world didn't transfer to the other.

But separate worlds are an illusion. I lived only one life, and inevitably those two worlds would crash and burn. For about three months, if either my wife or I would have had the slightest bit of extra time or money, we would have sought counseling and probably separated. She had good reasons to divorce me and that is what I expected. As it was, we hung on in a state of inertia driven by poverty and exhaustion, each of us uncertain, each of us sad, each of us lonely.

The walk home from the hospital took thirty to forty-five minutes, long enough for me to leave the emotional stress of work behind and think about home, wondering how long I would have that home. I would miss it greatly. My wife was wise and kind and I loved her. I loved my two sons more than life itself. And yet, I had blown the relationship. In a rare moment of self-awareness, I realized that the current crisis was the result not of circumstance but of character, and the same character flaws that had gotten me to this point would continue to follow me into future relationships. Furthermore—and this was the most damning realization of all—I could not fix it. I could not fix the relationship; I could not fix myself.

So I prayed, possibly the first truly sincere prayer I had prayed in years. I said, "I am so sorry. And I can't fix it."

That's all I said, to myself and the creator of the universe, and I still walked. I expected no reply. I had reached the New Haven green and crossed diagonally in front of the three empty, landmark churches in the dark and cold winter silence.

This is when Jesus spoke. "No, you can't fix it. But I can. Just go home."

The voice was real enough that I startled and looked

around. No other people were visible, of course not. The voice was "in my head," but it wasn't me talking; I know my own voice. This voice was not particularly warm. The tone was more of disappointment, maybe a bit stern. Resigned, maybe. And that's all He said. It really could only be Jesus. When you pray and somebody answers, who else could it possibly be?

I went home and told my wife that I would never leave her and never give her any additional reasons to leave me. I couldn't prevent her from leaving, but I wanted her to know that I did not want her to do that. I didn't begin with "I had a message from God" because I didn't think then, and don't think now, that that is a good approach to a difficult conversation. But maybe I ended with it, because I wanted her to have more reason for hope in the future than I had given her in the past. She, of course, did not believe me.

And nothing changed. Except that I felt more at peace. Whatever happened, God Himself had assured me He would fix it.

A few days later an issue arose about a time conflict involving getting the children to appointments. I don't remember the details, but I remember the conversation.

"I'll take care of it," I said, even though I had some reason to expect some work-related demand to come into play. I determined that I would find a way to fulfill this one family need.

My wife looked at me in that way that she has when she needs to remind someone that she is no fool. "I don't think I can trust you," she said.

Immediately I was angry. I formed the words in my mind: *What do you mean, you can't trust me? You've got to trust me! If we're ever going to make this thing work, you've got to trust me.*

But the words that came out of my mouth were: "I know. I haven't given you reason to trust me. But I will be here and do what needs to be done. And I will do it again. And again. For as long as it takes to earn back your trust."

She gave me the same I'm-no-fool look, took a beat and said, "We'll see."

I don't know what I looked like to her, but inwardly I was stunned. Those were certainly the right words to say, but not what I had intended when I opened my mouth. This was Jesus "fixing it." That moment was the first building block of a new relationship. Other blocks followed, and eventually other crises were faced together, other adventures, another child, adult children with spouses, grandchildren, and eventually a fiftieth wedding anniversary.

There is a kind of *before* and *after* thing following an experience like this. *Before*, it was okay to believe that God existed and created the universe and all and I should try to be nice and good whenever it is convenient and hope for the best. *After*, I realized that the God who created 100 billion galaxies or more is not so big that He does not pay attention to every thought, word, and deed that comes out of my puny little life and is standing by waiting for me to ask Him for help.

There's a bit of cognitive dissonance in the idea of a big God understanding and intervening for a small man. Maybe that's why God took the form of Jesus Christ with real flesh and real blood who walked on Earth. Two thousand years ago He talked to a ragtag bunch of Jews in the Roman province of Palestine, and He's been talking ever since to those who are willing—or driven—to listen. He did it in my life. And, in my *After* experiences, I have discovered that I am not the only one. For many of us oh-so-less-than-perfect people, Jesus has intervened and called us in very real ways, including through His words.

So, I believe in Jesus Christ, based on historical fact and our finest philosophy and the fulfillment of our collective consciousness. But mostly I believe because He talks to me and other sinners when we are still helpless and lost.

Chapter Five

"His only Son, our Lord"

In the beginning, God creates mankind in his own image. Then, with the birth of Jesus, God creates Himself in mankind's image. This has lovely literary and theological symmetry, but it is so difficult to understand.

Does the mind that conceived gravity, wrote the laws of thermodynamics and motion, structured quantum mechanics and relativity principles, designed the DNA molecule and the Higgs boson, now become confined to what a 1,400-gram human brain can know? Do the hands that once formed the Milky Way now become limited to the skills of a carpenter? Does the creator of the universe become re-created in the womb of a woman?

It seems unlikely.

But there are reasons to believe this unlikely truth. First, Jesus said so. Therefore, He either had a severe mental illness involving delusions of grandeur, or He was exactly who He said He was—namely, the human incarnation of the eternal and almighty God. Since His other teachings had internal consistency and wisdom rather than insanity, it is hard for any serious reader of the four accounts of His life to believe Him to suffer from a serious mental illness. He doesn't sound insane; He sounds like the Son of God.

The second reason to accept what Jesus said about Himself, is that three people who knew him well, and one person who talked to many of the people who knew Him well, wrote down what He said and what He did. And while all four of them differed in many details, they all agreed on one thing: Jesus is the Son of God. Since then, millions, possibly billions, of people have had experiences like them and like me. Jesus personally convinces us that He is the Son of God.

I can't think of another case in which a historical figure claimed to be the Son of God and the people who knew them best affirmed the claim. Certainly, there have been founders of great world religions, like Buddha and Mohammed, but they never claimed to be God. Certainly men have claimed to be god—various psychotics whom we currently heavily sedate and frequently confine to institutions, and megalomaniacs who happened to rule the Roman Empire or Nazi Germany or the Empire of Japan—but the people who knew them best didn't believe they were gods.

It is easy to believe something similar but distorted. If Jesus is the Son of God, and I am a child of God, does that mean Jesus is just another child of God? A kind of spiritual sibling? Brother Jesus?

I don't think so. If we recognize that God in the form of Jesus has taken on the limitations of humanity—that is, become the Son of God—but is still God in some difficult to understand cosmic sense, then He is also Lord.

Although that concept is not difficult to accept on paper, to live it out, for me, has been difficult.

In the middle of my first year in college I had no idea what to major in or what kind of job I would want after I finished college. Perhaps that was due to lack of exposure to the big world. My grandfather was the first in our lineage to finish elementary school and learn to read and write. My father was the first to finish high school. My generation would be the first to go to college. But other than the teachers at school, the preachers at church, and the doctor that we saw once a year, I didn't know any college graduates. Our neighborhood and my family were populated by mailmen and mill workers, cooks and clerks, policemen and firemen, carpenters and bricklayers, farmers and storekeepers. Yet, from my earliest memory I was expected to go to college.

Now, here I was in the middle of a legacy dream with no clear idea on how to proceed. So, one afternoon midway through my second quarter and before trying to register for third quarter classes, I lay down on my bed and asked the Lord to guide me. I figured since He wanted me to know He was real from my experience in the rain when I was fourteen, He might have some opinion on the direction I should go when I was eighteen. It surely was not obvious to me.

I got no answer, but I mused about who I was and what I was good at and what I liked. I faced the facts: I was pretty good at science but didn't love it, and I loved literature and the arts but didn't excel in them. I'd heard that medicine was an art and a science, so on that thin bit of reasoning, I got up and the next day registered for the first of my pre-med classes.

Was that guidance from the Lord or the musings of a

dumb eighteen-year-old kid? I don't know. At many points in my life I prayed for guidance and never got what I thought was a clear answer. I made the best decision I could at the time and moved on. That's pretty much the pattern of subsequent critical decisions in my life, and I always wondered if I was on the right path. I mean, what good is it to have a "Lord" if He doesn't tell you what to do?

Sure, He laid out principles in the Bible to follow. The Ten Commandments,[1] which, if I'm being totally honest, I have treated as the Ten Suggestions. The Beatitudes didn't resonate with me because I never understood how I could be blessed for being poor in spirit.[2] There's the Shema: "Love the LORD God with all your heart and with all your soul and with all your strength,"[3] but how does one do that? Of course, there's the golden rule: "Love your neighbor as yourself,"[4] but I defined a "neighbor" as someone I liked (clearly not Jesus' intention). So, I stumbled through life full of hypocrisy, imperfection, and uncertainty.

I suspected the noise of the world was drowning out the voice of the Lord, but I thought I was doing pretty good—not perfect but better than most and asking forgiveness in a general sense at each monthly Communion service. By claiming salvation by grace, I guess I fulfilled the minimum entry requirements to heaven. Coupling complacency with grace may still have gotten me into

---

1. See Exodus 20:1–17.
2. See Matthew 5:3–11.
3. Deuteronomy 6:5; see also Matthew 22:37, Mark 12:30, Luke 10:27.
4. Matthew 19:19, 22:39; Mark 12:31, 33; Luke 10:27; Romans 13:9; Galatians 5:14.

heaven, but it didn't let me hear the voice of the Son of God.

Then one afternoon I prayed an honest prayer, one I'm embarrassed to share: "Lord, I know I've been really sinful in the past, but I've put those things behind me and now I'm doing okay. As a matter of fact, I can't think of any serious sins I'm committing. So, Lord, show me my sin."

Now that's a prayer that is as arrogant as it is honest, and maybe the only reason I dared pray it was that, based on past experience, I wasn't expecting an immediate literal answer.

"You're greedy," said the Voice in my head that was not my own.

"How can You say that?" I replied with amazing temerity considering who I must have been talking to. "I give to multiple charities and tithe to my church."

"You do that because you can afford it."

Well, granted. By this time my income was such that we lacked for nothing even after what I considered my generosity. But I didn't want to tell Him that.

The Voice continued. "You were invited to a mission trip, and you didn't go because of greed."

"That wasn't the reason," I said. "I could afford the $2,500. I didn't do it because I couldn't leave my practice for the sixteen days the trip would take."

"You don't *want* to leave your practice because you are afraid of losing three weeks' worth of billing."

And this was true. It was more than simply giving up my paycheck for three weeks. If I didn't work for three weeks, I wouldn't bill for the income that paid for my employees' salaries, the office rent, and insurance in its various forms. Expenses would continue while income dropped to nothing. It was potentially a lot of money and was truly the reason I never even considered going on this mission trip.

"It's past the deadline to sign up," I said, but by now the argument was lost. I was making excuses.

"Make the call," said the Voice.

So, I made the call and went on the mission trip and became a little more humble and a little more generous. If I became poorer as a result, I haven't noticed. What I noticed since then were the moments—sometimes on mission trips, sometimes in the operating room, sometimes sitting with a friend or even a stranger—when despite whatever physical discomforts or dangers or embarrassment I faced, I had the pure knowledge and joy that I was doing exactly what I was created to do.

And that's the payback. If I choose to live as a child of God, I am a being with purpose and destiny—if I allow Him to be my Lord.

He *is* the Son of God, but we *choose* to let Him be Lord.

Chapter Six

"Who was conceived by the Holy Spirit and born of the virgin Mary"

I don't like to think about this part of the Apostles' Creed. I tend to mumble through it, figuring I'd only have to care if I was Roman Catholic. I don't understand it. This is technically the first in vitro fertilization (IVF) except it was in vivo and the donor was divine. The divine union of God's DNA with that of a Jewish teenager? Really?

I expressed my skepticism to Dr. Jeff Hoy once while we were on a trip to Israel. He listened and calmly said, "The biggest leap of faith has to be the resurrection. If you can believe that a dead body, days after decomposition, could rise from the grave, then the virgin birth should be no more difficult."

Touché. Some things are to be taken on faith. But things taken on faith should be plausible in some way and important. Two historians of Jesus' life, Matthew and Luke, insisted on a birth story that included a virgin birth. They had to believe it was true after speaking with the mother of Jesus and others who were still alive when they were writing. And they had to believe it was important.

It *is* important. Why? The virgin birth cements the reality that Jesus is all-in as a human being. Sure, He healed the sick and raised the dead and walked on water, showing the world that He had divine power, but He also limited Himself in critical ways. He could not or would not use his divine power to intervene for Himself. Born as a baby, He wore diapers and pooped in them. He worked, perspired, grew weary, and slept. As an adult, He went to parties that served alcohol; He *supplied* the booze at a wedding. He laughed. He cried. And He did not get out of this life without hunger, loneliness, beatings, whippings, and, ultimately, torture and death.

The humanity of Jesus is deeply important theologically, not least because it gives the proper conditions for His ultimate sacrifice to save mankind from sin. But, to me, this humanity to save the sinners is just as vital to comfort the innocent. This is the answer to Job when he cried out to God demanding a reason for his suffering.

The only purely innocent person in the Bible is Jesus. Job is nearly as innocent. He suffers greatly for reasons that are beyond his control. His story, I am told, is thought to be the most ancient of all our biblical texts and it deals with the perplexing question of innocent suffering: Why does a good God allow innocent people to suffer?

Job is long and poetic and has a dramatic conclusion—totally worth reading and thinking about—but I don't think it hits home until you've experienced innocent suffering yourself, and I don't think the answer to Job is complete until we understand Jesus as "conceived by the Holy Spirit and born of the virgin Mary."

I was the junior resident in neurosurgery when I got a call to the pediatric emergency room. A four-year-old boy had been brought in by ambulance from home because he was unresponsive. He had been resuscitated and intubated before I walked into the room. The first twenty seconds at the bedside told me he had been beaten over every square inch of his body. The next two minutes of examination told he was already brain-dead.

Innocent suffering.

A twenty-two-year-old woman was driving home one Friday night, and a drunk young man was following her, tailgating and speeding—his apparent method of flirtation. He rear-ended her car, and she was left paraplegic for life.

Innocent suffering.

An eight-year-old boy, balancing on a curb while waiting for his school bus, fell into the street in front of a truck. He arrived in the ER, severely injured and in deep coma but not quite brain-dead. Our team struggled all day to bring him back, but by evening it was clear that we had lost; the child died.

Innocent suffering.

The parade of innocent suffering and grief started before Job, and it hasn't ended yet. You all have your own stories. I've heard some of them. Despair would not be unreasonable. Job himself cried out in pain and indignation before God answered him out of the whirlwind.[1]

Then along came Jesus. He got whipped, beaten, and hung naked to die a slow, painful death. He was not a distant spirit in the sky professing sympathy for the human race. He was not a universe creator who set the ball rolling a few billion years ago and let nature take its course. He was not the celestial judge who stands above it all and punishes us for sins real and imagined.

---

1. See Job 38:1.

No, this was God who tied Himself to humanity literally by His DNA. He is eternal, meaning His existence is not bound by time. Two thousand years ago He felt nails driven into his wrists. Today, He feels what a rape victim feels, what parents feel when their child suffers, what an old man feels when he buries the love of his life. He stands in the concentration camps and sits in the cancer wards. He starves with the hungry and He takes the bullet from the firing squad. Because this is how God answers Job. This is how God loves the world: He becomes Jesus.

The Bible tells us "Surely he took up our pain and bore our suffering."[2] Since He is eternal, He takes up the pain and suffering of every human being, before and since His earthly lifetime. This means that in some cosmic sense, when that child got hit by a truck, Jesus got hit by a truck. When the child's parents grieved, Jesus wept. He takes up our pain and bears our suffering. When Job lay diseased on the dung heap longing for his lost children, Jesus lay with him.

So, I don't know the details about the Holy Spirit and the virgin Mary and how that works with sperm and ova. The details are neither provable nor understandable. What I do understand is that God irrevocably bound Himself to the virgin Mary to be born as a helpless infant into a lost world. He is human. He has our DNA. He celebrates with us, He suffers with us, He dies with us. And we can be resurrected with Him.

This is how God loves the world.

2. Isaiah 53:4.

Chapter Seven

"He suffered under Pontius Pilate, was crucified, died, and was buried"

God is dead. Not only is He dead, but Pontius Pilate, a specific historical figure, marks the event. He is dead by crucifixion, which by Mosaic law puts Jesus under God's curse.[1] There is no mistake: His body is dead and buried.

It was true then, and there are times now when it seems like He is dead. I don't see Him, I don't hear Him, I don't feel Him. Day after day, week after week, month after month, I do the best I can, but the world is a dreary place. One of those times occurred while I was a neurosurgery resident during a particularly difficult season.

I was called to the ER at Yale-New Haven Hospital by stat page and found the trauma room full of nurses, X-ray techs, ER doctors, the general surgery resident, all resuscitating a twenty-something-year-old kid from a motorcycle accident. Details filtered in as his stomach was pumped (remains of a recently consumed Italian dinner and much alcohol), an endotracheal tube placed, X-rays of

1. See Deuteronomy 21:22–23.

his chest and skull were taken, and we waited for the CT scanner to be available. He had ridden his motorcycle at full speed down a hill in West Haven, gone through the T-intersection, directly striking the telephone pole on the other side. No helmet, little hope.

All the residents at our hospital did rotations at the West Haven VA. We had all driven that same road dozens of times on our way from here to there and back again; we all knew exactly what hill he had accelerated down, what telephone pole he had hit. We all envisioned it; we all knew it was insane, even beyond the usual testosterone and alcohol-induced risk-taking behavior of young men.

The name was familiar to old hands in the ER, and they called for records. Another motorcycle accident had occurred at exactly the same place on the same date two years before, and the victim had the same last name. As in the current case, the victim had arrived in the ER and died shortly thereafter. That victim was the brother of tonight's victim. From this I created a narrative in my mind. Maybe we all did, but nobody talked about it. It was too sad.

A troubled young man, depressed over something—a lost girlfriend, a lost job, a failed test—decided to self-medicate with alcohol. Perhaps he had an underlying mental health disorder. He had a mom who cooked him dinner. Maybe he had a dad and he certainly had a younger brother. But it wasn't enough. The future was too dark. He rode into the night, accelerating his motorcycle to maximum speed on a downhill stretch and ended it all on a telephone pole. Two years later, the little brother faces a loss that seems big and sees his brother's solution. So he takes his motorcycle on the same path to the same pole. There's a mom left behind still washing the dinner dishes. Maybe a dad.

After the resuscitation in the ER, we took him to the Neuro ICU on a respirator with an intracranial pressure monitor. When we got him settled there, I noticed the tattoo on his left deltoid: "Born to Lose." A few hours later, he was dead.

A couple of weeks later, a sixteen-year-old girl on her way home from school, suddenly cries out in pain, clutches her head and collapses. Rescue brings her to the ER in a coma. A CT scan shows a large hemorrhage in her left hemisphere, the part of the brain that controls speech, language, and movement of the right face, arm, and leg. The bleed had been caused by something called an arteriovenous malformation (AVM, for short), something that would have been a formidable challenge for an experienced neurosurgical team with weeks of preparation. But she needed immediate surgery. So she got me, a resident, and a junior staff neurosurgeon.

The operation was long and difficult. Though we *knew* a lot about AVMs, we both had minimal *experience* for this condition. The urgently gathered OR staff was also inexperienced. For every clot removed, an artery bled. Whether this artery could be cauterized without permanent loss of speech, coordination, or intellect, we had no way of knowing. If abnormal blood vessels were left uncauterized, the girl would re-bleed and die. My held breaths alternated with hyperventilation. Hours passed.

Finally, all the clots were out, the AVM resected, and the brain appeared normal. We closed. After surgery, she awoke without speech or comprehension, paralyzed on her right side—a life saved but at the cost of probable severe lifelong disability. We didn't celebrate. Two weeks later, her condition unchanged, she transferred to a rehabilitation facility to recover with people three and four times her age.

A few weeks later, an eight-year-old boy was playing on the sidewalk waiting for his school bus on busy Whitney Avenue. He tripped and fell into the street in front of a garbage truck. Brakes screeched, horns blew, but still that sickening thump. Traffic and time stopped. An ambulance transported the boy to the ER on a respirator in deep coma from the head injury.

He got hit in front of the preschool my son attended. I called the same junior staff neurosurgeon and described the case. He asked my assessment, and I told him it looked hopeless. Well, almost hopeless. We gambled with the *almost* and ignored the *hopeless.*

I placed an intracranial pressure (ICP) monitor to guide treatment, optimized the respirator rate, and administered various medications to control brain swelling. By 2 p.m., these measures had failed. Another CT showed diffuse brain swelling. At 3 p.m. we took him to surgery to remove a large part of his skull so that even as the swelling increased, his brain would not be compressed and the damage would be limited. But even as the skin sutures were placed, his brain continued to swell. The ICP rose again to dangerous levels, and then to levels incompatible with life. By 6 p.m., the battle was clearly lost. The child would die.

It felt like God was dead. I had been part of surgeries that had failed, and this was not the first person I had seen die, nor even the first child. But this was hard, almost too hard. Maybe because this time the accident took place in my neighborhood and the victim looked too much like one of my own children.

Maybe everyone has times that it feels like God is dead. I imagine how Jesus' family and disciples felt on the Saturday following Good Friday. It not only *seemed* like God was dead; He *was* dead. And buried. The future held nothing except futile struggle and inevitable despair.

That's how I felt. So I hid behind the paperwork in a corner of the ICU until the staff neurosurgeon came and stood across the desk from me. "There's someone to see you," he said.

I shook my head; if I tried to speak I might cry. Besides, no one ever came to see me; I was just the resident. Or worse, maybe he wanted me to talk to the parents. I couldn't face them. Not yet.

"Come on," he said.

"No," I managed to say.

He stared me down, long enough that he knew I was in pain, and I knew he wasn't going away. "Come on," he said again.

I got up and followed him. After all, he was my boss.

In the hallway outside the ICU stood a teenage girl in a cowboy hat. Long, dark hair spilled out from under the right side of the hat, but no hair on the left, and maybe a hint of a scar in front of her left ear. She smiled a crooked smile and leaned on an aluminum four-poster cane.

Then I recognized her. The teenager with the AVM. Not in a coma, not a vegetable. Walking and ... could she be talking?

Her words were slow and overly round like her tongue was too big. "Thank you," she said.

I don't know what I said, and yeah, I might have shed a few well-disguised tears of joy.

Maybe it was a coincidence that this thoughtful teenager took this particular moment when my boss and I were in the depths of futility and despair to say thank you. But *coincidence* is just another way of saying *miracle*.

Each and every one of us, at one time or another, feels like He is dead. We are living out the Saturday after Good Friday, and it is a long day and a longer night and the darkness feels like it will never end. There is a tomb in a cave not far away with a big rock closing the entrance, and it is beginning to smell bad.

But morning is coming. Easter morning. Dawn breaks over a spring morning, the rock is rolled away, light comes into dark places, and the tomb that just moments before held despair and decay—is empty.

Chapter Eight

"He descended into hell"

A few years ago I took my canoe and my somewhat hesitant adult son, Adam, on a Suwannee River canoe trip from the Okefenokee Swamp to White Springs, Florida. The water levels were so high that we launched from the parking lot because the boat ramp was submerged. The first day we paddled through the lower branches of blooming tupelo trees whose trunks were underwater. Although this would be a beautiful and unique trip, I knew from the beginning that on the third day we would come to that part of the Suwannee known as Big Shoals, and I wanted nothing to do with it. We were paddling my beloved cedar-strip, hand-built canoe, which is the best flat-water canoe ever, but it is not built for rapids. It is breakable.

The morning of the third day, I took care to lash all loose equipment to the gunwales or the scuppers, we both tightened up our PFDs, and paddled slowly looking for the signs that I had been told would warn us when we were near the shoals and direct me to the portage site. Bluffs on either side of the river got steeper and I figured we must be getting close. I paddled slower and looked carefully on both banks for the portage. Then I heard a low-pitched growl coming from downstream. I stopped paddling and listened. The growl became a roar, and I knew we were about to be drawn into the rapids.

Unbeknownst to me at the time, the water levels had gotten so high that the signs and docks and stairs that marked the approaches and the portage were submerged, and the rapids were running as high as they ever would. Maybe the portage itself was submerged; I never got to find out.

For a brief moment we could have turned the canoe and paddled upstream as fast as possible. We might have made it and found safe passage. But the moment was fleeting, and I thought: *How big could Big Shoals be in a state as flat as Florida?* And *I've run rapids before; how hard could it be?*

So instead of turning around I thought, *To hell with it*, and pointed the canoe downstream paddling briskly into the maelstrom.

Big Shoals may not be Niagara Falls, but it sure felt like it. The low-pitched growl had turned into a deafening roar as the bow dropped at a 45-degree angle into white water. I steered around a coming boulder—all good, missed the boulder—but the bow (and Adam) went into a deep trough and never came back up. We were sunk. The canoe, packs, paddles, and our hats were driven through the next two levels of rapids as we hung on to the canoe for dear life. Finally, we found ourselves floating in flat water again, and I literally put the bowline in my teeth and swam for shore.

We survived along with the canoe and all our equipment. Even the water bottle that had floated off appeared twenty minutes after we got on the river again. Getting soaked was perhaps a pretty good trade-off for a good story. As usually happens when I say "To hell with it," I abandoned a well-thought-out plan for a risky alternative and bad things happened.

At the onset of Jesus' public ministry, the devil tempted Him in the wilderness. He offered Jesus the opportunities to use His powers for self-fulfillment, to use His powers for political success, or to use His powers for self-preservation and celebrity. Jesus, as we know, turned these options down. He did *not* say, "To hell with it." He followed His Father's original plan.[1]

What did the devil do? "When the devil had finished all this tempting, he left him until an opportune time."[2] When was this more opportune time? I think in Gethsemane just before His arrest, trials, and crucifixion. He appealed to the Father in heaven: "Father," He said, "everything is possible for you. Take this cup from me." The human Jesus knew what was coming unless divine intervention found another way. But He continued the prayer: "Yet not what I will, but what you will."[3]

In the completion of His prayer, He reversed the decision that Adam and Eve made in the other garden, the garden of Eden, when they decided to step back from God and eat of the fruit of the tree of Knowledge of Good and Evil, a decision influenced by the same devil. Adam and Eve decided to follow their own will; Jesus committed to follow the Father's will, even though it meant torture, humiliation, death. And, yes—hell.

---

1. See Luke 4:1–12.
2. Luke 4:13.
3. Mark 14:36.

His last words on the cross are, "My God, my God, why have you forsaken me?"[4] Then He dies and goes to hell, into chaos and destruction, alienated from His Father in heaven. "To hell with it" takes on a whole new meaning.

Sometimes I say, "I went through hell for that," but I don't really mean *hell*. I mean it was hard work with mental and physical suffering. The real hell will also include isolation and alienation from all good things, our loved ones, and the Father in heaven.

Our family still refers to 1991 as The Bad Year—not hell exactly, but close enough. My wife was diagnosed with breast cancer and underwent surgery and chemotherapy. Before her last chemotherapy appointment, my son Adam (of Big Shoals adventure fame) was diagnosed with a brain tumor and treated by biopsy, shunt, and radiation. He subsequently also suffered with complications from a shunt infection. Then, equally life-threatening and terrifying, my other son was robbed at a neighborhood mall by a man wielding a shotgun.

I know some of your stories, and many of you have gone through things much worse. I expect that most of you know what it is like to lose someone or to lose a dream or both, and you have experienced the darkness and silence that follows a loss. There are no right answers for how to cope. Everyone follows their own path.

Our family kept going to church. We kept praying. We kept believing. But we felt forsaken. Like Job; like Jesus. For a long time, nothing but silence came from above. I knew He was there listening, but if He was talking, I couldn't hear Him. A wise family would have gotten

4. Mark 15:34.

counseling, but we just muddled along. Sometimes you are in too deep to see alternatives. Each of us dealt with our own experiences of depression and that emotional realization that the world was not a safe place. The unthinkable not only *might* happen, it *did* happen, and will likely happen again. The best we can hope for is to enjoy the moment before the recurrence of sickness or death.

Cancer recurrences did not come, however, and Mary and Adam recovered physically and emotionally. The process wasn't easy, and each of them has a story to tell and scars to show. My scars were invisible, but I was chronically depressed about the futility of existence and the ultimate death of everything and everyone. The malaise was pervasive and persistent.

Five years after our Bad Year, on a Good Friday afternoon, I sat alone in our backyard. Sadness filled me, and I noted that this was the day that even Christians celebrated death. The certainties I knew beyond death were all infinite: time, space, matter. Then I experienced what I can only describe as a vision of the incredibly complex mind of God, and these great infinities were but thoughts in that mind. The mind of God has places for Mary and Adam and all the people I love. The mind of God encompasses the great and the infinite and the eternal, but it also includes the tiny and the finite and the mortal. It includes me.

It is hard to imagine the experience of Jesus between His death on Friday afternoon and His resurrection on Easter morning. The text tells us that He experienced being "forsaken" by God, helpless and hearing only silence from above. There is something endearing about this, that God takes human form in such depth that He can feel isolation from all the things that make Him God: love, power, strength. He knows from first-hand experience what it is like to be in hell.

Easter Sunday morning came thirty-six hours after my Good Friday in the backyard. The long silence from above was over. I pray that each of you who now feels that silence will find your own Easter soon. Jesus feels what you feel. He sees you; He hears you. He will speak to you if you will listen.

Chapter Nine

"The third day he rose again from the dead"

My friend Jimmy Herman (a pseudonym), the thief, told me about meeting the risen Jesus.

Jimmy and his brother lived across the street from me in a low-rent neighborhood scheduled for demolition by the end of the summer. He had been recently released from prison and worked the late shift as a short-order cook at a greasy spoon on Skid Row. He was trying his best to turn his hardscrabble life around. I worked nights in the hospital lab, so we were on a similar off-beat schedule.

I met Jimmy after his brother had a psychotic break. Jimmy was not home when his brother started roving through the neighborhood, ranting incoherently. My roommates and I calmed him the best we could and took him to the ER at the county hospital. Unfortunately, in the parking lot, before we could get him registered, he assaulted a police officer and landed in the city jail. The next morning I came up with the fifty bucks to post bail.

We all came from communities where neighbors took care of neighbors and naïvely transferred our habits to the inner city. Jimmy had a different background. He had been crippled by polio as a child and either orphaned or abandoned. He had been raised by his grandmother in a tough neighborhood and learned early how to fight with his crutches. By the time he was an adult, he had beaten up so many people that he carried a permanent restraining order preventing him from using crutches. At thirty he was a career criminal and self-described thief. When we met him, he had just finished his third term at the state penitentiary.

After the psychotic break and the night in jail, Jimmy's brother disappeared, hopefully to a treatment facility. The rest of their family were dead. Many of Jimmy's other friends were in prison, and those who were out were actively engaged in crime. Jimmy was looking for a way forward that did not involve yet another stay in the state penitentiary. I guess he was lonely. Our kindness to his brother had surprised him. He started coming over frequently, sharing meals, sharing stories.

One of his stories was about the crime that led to his last stint in prison. Though usually a simple burglar, he and some associates devised a plan to rob a bar on their busiest night. One man stood lookout on the roof, another blocked the entrance on the street, and three masked men went into the bar to terrorize the patrons and empty the cash register. Jimmy, with leg braces from childhood polio, did not have the mobility to be the roof lookout or agile enough to be inside the bar, so he blocked the entrance on the street.

Of course, the police came. This robbery was a copycat plan from other recent crimes, and the city was prepared to shut down this kind of crime with multiple vehicles and officers ready and waiting only for the silent alarm. A gunfight ensued as Jimmy and his friends tried to flee. The rooftop observer was killed, and Jimmy was gravely wounded. He remembered the impact of the bullet and falling to the sidewalk. Just before losing consciousness, he realized that the dark sticky puddle under his face was his own blood.

He described a deep darkness followed by a tunnel of light, a figure in the light—Jesus, he knew, though he could not see His face. Next to this figure sat his grandmother, the only person who had ever loved and cared for him, her hands out to Jimmy, beckoning to him. The scene was so soft, so warm, so inviting, he wanted to join her in the light. But he was also afraid. He knew he wasn't worthy to encounter Jesus. He had just shot at a policeman, maybe killed somebody. If he took the hand of his grandmother and the hand of Jesus, hell might await him. He could never come back to this world. Nobody had ever convinced him of the doctrine of grace, so he fought to stay in the darkness. He awoke Sunday morning handcuffed to a bed in the county hospital. He faced a stark reality in this world; yet, in the world to come, he saw a new reality of hope.

That Jesus rose from the dead is the most difficult tenet of the Apostles' Creed to accept intellectually and the most important tenet to take on faith. If you believe it, you are a Christian; if you do not, you are not.

No one doubts that Jesus died. The Gospel narratives and the historical commentators agree on that, and we do too. Everyone dies, why not Jesus? But no one comes back from the dead. Not really. That is the stuff of myths, legends, and zombie movie plotlines.

The Bible gives several other accounts of people returning from the dead. Lazarus is the best-known example,[1] but there is also the son of the widow of Nain,[2] and the daughter of a synagogue leader.[3] These survivors appear to have gone on to a life on earth and eventually an earthly death. These are resuscitations rather than resurrections.

Some others may not have experienced earthly death: Enoch, who "was no more, because God took him away,"[4] and Elijah, who was separated from his companion by a chariot of fire and went up to heaven in a whirlwind.[5] But neither of these men ever again walked the Earth and interacted with living men.

Only Jesus and His followers make the claim of *resurrection*, which means return from the dead with no implication of an endpoint to the new life. St. Paul reports five hundred witnesses.[6] Jesus returned to life on earth and walked and talked with other living humans. No figure in history or religion makes the same claim. Of the estimated 109 billion humans who have ever been born and died, Jesus is the only exception. Could it be true?

I have heard alternative explanations from people who like the ideals of Christianity but can't believe in the resurrection. One explanation is a Passover plot to substitute a Jesus look-alike for the arrest and crucifixion, then switch the bodies on Easter morning. Another explanation has the disciples resuscitating a drugged and injured —but not dead—Jesus in the tomb after the crucifixion for

1. See John 11:43–44.
2. See Luke 7:14–15.
3. See Matthew 9:18–26.
4. Genesis 5:24.
5. See 2 Kings 2:11.
6. See 1 Corinthians 15:6.

the Easter morning reveal. Or the simplest scam: Steal the body and spread rumors of Jesus appearances.

The more modern denial of the resurrection takes the form of a virtual resurrection. In other words, perhaps Jesus' followers believed His ideas were so powerful that they deserved to be formed into a new religion, so they invented the Easter resurrection myth even though they knew He was dead.

But if we apply Occam's razor—namely, that all things being equal, the simplest explanation is most likely true—we have more evidence that the resurrection of Jesus is true than any of these other explanations, or, for that matter, any other event in first-century history.

The Bible presents us with two immutable facts. First, the Gospels are candid, unpolished reports from four different sources who disagree on many details, but all agree on one central fact: Jesus returned from the dead. Second, none of the eleven apostles who lived past that first Easter morning preached anything except real, physical resurrection, and all but one were martyred for their faith. This is a good reason to believe the unbelievable. Eleven men gave up their fortunes, families, and professions to tell about the resurrection of Jesus even though they faced imprisonment, beatings, whippings, and eventually death. I might believe one fanatic would live and die to perpetuate a scam, but would these eleven men, several women, and other believers all sacrifice their worldly lives for a *virtual* resurrection?

We believe because they believed, and they believed because they saw Jesus. The resurrected Jesus keeps appearing—not merely as a misty inspiration for generous and righteous living but as a tangible fact in the lives of believers and even nonbelievers. Jimmy's experience is not unique. I have heard stories from my generation and previous generations, usually told quietly with awe to close family and trusted friends, because they are precious. Some of you have your own resurrection stories and Jesus sightings. They are stories told at campfires, in prayer rooms, and recounted at funerals and weddings and baptisms. These are testimonies to the resurrection of the living Christ that have been going on for 2,000 years. The resurrection of Jesus was a historical event. But it is also eternal—beyond the restriction of time—occurring every day in encounters that are intensely personal.

This is why we believe. Jesus is still alive, showing up even for the most unrepentant sinner at the time of his greatest need. He showed up for St. Paul on the road to Damascus.[7] He showed up for Jimmy on a bloody sidewalk in Minneapolis. He showed up for me on the New Haven green. He rose from the dead, and He will show up for you.

---

7. See Acts 9:1–31.

Chapter Ten

"He ascended to heaven and is seated at the right hand of God the Father almighty"

My father was a deacon in our Lutheran church in Minneapolis when I was a teenager and young adult. This meant he wore black robes once a month and helped to serve Communion. He also went to a lot of meetings and knew all the pastors on a first name basis. It involved a lot more, but I was privy to very little of it. My dad was fairly laconic, and truthfully I wasn't interested.

Until one Sunday afternoon when he took a call in my room while I was a college student. Back in the day of landlines and short cords, there were only limited places one could talk on the phone, and on that day my upstairs room was the only available quiet place. He ignored me while I studied at my desk on the far side of the room, but I could not ignore him.

A kid from the confirmation class had apparently been given the assignment to ask a deacon a question about faith. This junior high boy called Dad to ask what heaven was like.

Although I was only getting one side of the conversation, I got that much, stopped poring over the extremely puzzling calculus textbook, and started listening. I suppose Dad could have told him about the streets paved with gold[1] and the river of life lined by the twelve crops of fruit of the tree of life as outlined in the revelation of John.[2] But he didn't. He told the boy he didn't know what it would be like physically, didn't know about wings or no wings, harps or no harps, clouds or mansions or any of the other common metaphors. He told him that his best guess was that it would be a feeling like the one that comes at the end of a day of hard work when you know you've accomplished something worthwhile and you know that everyone you love will have enough food and enough clothing and a safe, warm place to sleep that night, and everything is right with the world. Dad told him that sense of joy and peace didn't come very often, but it was his best guess about heaven. It would feel like that.

Then Dad listened for a long time. He shared an occasional word of comfort, but his back was to me and his voice was soft. I waited, staring down at the still incomprehensible book, my mind now a million miles away from calculus. When he got off the phone, Dad quietly explained the conversation to me. The part I hadn't heard was that the boy had lost his father six months before. His question wasn't academic, it wasn't theoretical, it wasn't theological. It was personal. The boy wanted to know, *Where's my dad now?*

It is hard to imagine heaven. We can quickly let go of the amusing images of winged angels strumming harps while seated on clouds, and the long-bearded, white-

---

1. See Revelation 21:21.
2. See Revelation 22:1–2.

robed St. Peter checking the big book for reservations at the pearly gates. In the Creed, the word *ascended* loses literal meaning in our current understanding of the physical universe. Ascending takes one ultimately to outer space, not heaven. And the phrase cannot literally mean that Jesus is "seated." Why not standing? Why not flying? And what can "the right hand of God" possibly look like? Not four big fingers and an opposable thumb, surely.

We are reduced to metaphor because heaven is beyond our mortal comprehension. Ascended makes sense in that heaven is separated from the dirt and the darkness that we find when we descend. Heaven is somewhere, but it seems to be in another sphere than our current space-time continuum. Jesus may or may not be seated but He is surely in the presence of God Almighty and united with the Father in a way that is so far beyond our experience that we can only term it "at the right hand."

My dad wasn't much of a music fan or art critic, but after he moved to Florida at the age of seventy he had a favorite print on the wall of his townhouse. It was a rural landscape of snowy fields and barren trees, just past sunset with a golden horizon below and an indigo sky above. In one corner, small compared to the scale of the painting, sat a farmhouse with a wisp of smoke coming from the chimney and warm light pouring from the windows. To everybody else, the print looked like winter on the farm. To Dad, it looked like heaven.

I don't know about whether Jesus ascended or got beamed up or got transformed. I don't know if He's sitting or standing or flying. I don't know what the right hand of the Father almighty looks like. But I'm sure that Jesus is united with the Father in some way I can't comprehend, except that the union is the ultimate source of all peace and all joy. I won't be lonely. Jesus is there. So is Dad.

Chapter Eleven

"From there he will come to judge the living and the dead"

Once I did a good thing, and I wonder if it is enough to save me. The Sulzbacher Center for the homeless had just opened in Jacksonville, and the county medical society was trying to find volunteers to staff a free clinic there on Saturday mornings. They were desperate enough to use a neurosurgeon, so I showed up and did my best to be a decent primary care provider for a few hours. Among the people I saw was a healthy-looking Black man about my age and my size who had hypertension and needed his medication renewed. The other problem I noticed when I examined him was that his pants didn't come close to fitting. He had left the zipper undone so he could hitch them up over his hips then left his shirt untucked so that maybe no one would notice.

I liked the guy. If I got the connection right, he was the guy the cops knew as "Ten-Speed," a moniker he had picked up because the only thing he owned when he started dealing was a ten-speed bike. He built a little business and had a house and a car by the time he got busted.

Now, after some jail time, he was back on the streets minus the house, the car, and even the bicycle. Or pants. It's hard to get work without pants.

I heard other hard luck stories that morning. If you've ever volunteered at a homeless clinic, you know there are lots of sad stories and seemingly impossible situations. Generally, the best policy is to keep your head down, do your job diligently and without judgment, and let the rest go. You can't fix everything.

But somehow, *no pants* bothered me. I went home to my closet, found a couple pairs of very serviceable pants, a belt that could make it work if the pants were too big, a Bible, and a satchel to carry it all. I drove back downtown and asked around the homeless center until I found the guy and gave it all to him. Yes, he was grateful. And I'm grateful to him, because it makes it a bit easier to face Jesus as the judge.

I don't like to think about Jesus as "The Judge." He is our Savior. He loves the unlovable and forgives the unforgivable. How can He be the same guy that brings the hammer down? When I envision the judge in heaven, I relapse into the vision of the bearded old man on the throne sending the righteous to their mansions of eternal bliss and the unrighteous to everlasting hellfire and damnation.

But Jesus tells us clearly that He will be the judge, and the righteousness standard is unexpected. He says, "When the Son of Man comes in his glory, and all the angels with him, he will sit on his glorious throne. All the nations will be gathered before him, and he will separate the people one from another as a shepherd separates the sheep from the goats."[1]

He says to the sheep, the ones on His right: "Come, you who are blessed by my Father; take your inheritance, the kingdom prepared for you since the creation of the world."[2]

That sounds good. I like this part of judgment. Then, He goes on to give the reason for their eternal reward: They fed him when He was hungry, gave Him something to drink when He was thirsty, invited Him in when He was a stranger, clothed Him when He was naked, looked after Him when He was sick, and visited Him when He was in prison.[3]

Interesting.

He didn't list regular church attendance, saying "the sinner's prayer," receiving Last Communion, or even forgiveness of sins. Neither did He list financial contributions, large or small; a well-ordered, thoughtful life; nor efficient running of large charities. Nope. The sheep getting past the pearly gates sounded like kind and generous people who did little things, one at a time, for Jesus.

But they didn't do it for Jesus; at least they didn't think so. The sheep said, "Who, me? I don't remember that."[4]

Of course they don't remember doing that. Jesus ascended to heaven back around 33 AD. Not likely any of us are going to find Him in jail looking dehydrated,

---

1. Matthew 25:31–32.
2. Matthew 25:34.
3. Matthew 25:35–36 (paraphrased).
4. See Matthew 25:37–39.

scrawny, sickly, or naked. Even if He did come back and chose that place and appearance, would the sheep recognize Him? I wouldn't.

Fortunately, Jesus made it clear to the sheep (and to us): "Whatever you did for one of the least of these ..., you did for me."[5]

Whoa! I thought I was going to heaven because I said I believed Jesus was the Son of God and He forgives my sins. Now I find out that I need to complete the checklist of caring for the hungry, thirsty, lonely, naked, sick, and imprisoned? I'm terrified, my hope now hanging on the one time I shared pants with an ex-con.

And that's the good part of Judgment Day. It gets worse.

The Son of God is still on the throne. Now He is directing the goats to "eternal fire prepared for the devil and his angels."[6] Because they didn't feed the hungry Jesus, give water to the thirsty Jesus, take in the stranger Jesus, clothe the naked Jesus, care for the sick Jesus, or visit the prisoner Jesus.[7]

Not because they were murderers, rapists, thieves, adulterers, or liars. Not because they claimed to be atheists or agnostics. Nope, their sin was one of omission. They hadn't taken care of the down and dirty Jesus.

The goats likewise answer, "Who me? I don't remember that."[8] Maybe they also asked about grace. Maybe they asked for a bit of credit for finishing the Bible study course and memorizing a lot of Scripture. Maybe

5. Matthew 25:40.
6. Matthew 25:41.
7. Matthew 25:42–43.
8. Matthew 25:44 (paraphrased).

they pointed to evangelism successes. Maybe they claimed their baptism, Communion, and last rites.

But no such luck. "Whatever you have *not* done for the least of these, you did not do for me,"[9] He answers.

The goats are going to burn in hell.

Everything about the words of Jesus in the twenty-fifth chapter of Matthew is disturbing. First of all, hell is real. I like to believe the modern myth that the righteous go to heaven and the others go to sleep. This is the illusion that allows for suicide and euthanasia. I don't like to think of "eternal fire prepared for the devil and his angels." Although details are sparse and no doubt beyond our imagination, hell is not a good ending. And now we find out that we get there by being not necessarily bad but simply uncharitable.

Charity is good, sure. But who would have guessed that eternal joy or eternal suffering would hinge on charity? And a very specific kind of charity at that. It is *unconscious* charity that is eternally rewarded, and it is performed for "the least of these" regardless of who they are or whether they are complicit in their misery. In other words, they are both the deserving and the undeserving poor, they are the ones who got sick and the ones who made themselves sick (think alcoholics and tobacco addicts), the stripped and the strippers, the shunned and the hermits. They are the innocently imprisoned and the career criminals.

This is a narrow path. You have to be the kind of person who sees suffering and can't stand to see suffering. You have to ignore all the good reasons for doing nothing: "It won't matter, anyway." "They've gotten themselves

9. Matthew 25:45 (paraphrased).

into this mess, they'll have to get out of it." "I'm not equipped." "I can't afford it." Then you must expect no recognition and expect no reward. You must be *that kind of person*. The *only* thing that saves you is becoming *that kind of person*.

I'm not that kind of person. Not really; not all the time. Maybe nobody is. I'm wondering if my one act of kindness is enough to get me in with the sheep, but I don't think so. I have to rely on that other thing called grace.

Jesus also said, "I am the resurrection and the life. The one who believes in me will live, even though they die; and whoever lives by believing in me will never die."[10] That's "saved by grace," but I found out it doesn't mean just saying the magic words "I believe," and expecting the free ticket to heaven. Believing in Jesus means surrendering my will to His, listening to Him through prayers and the Bible, and actively seeking Him in my life experience.

Sometimes that life experience can be pretty damn short. Think about the thief being crucified with Jesus and the promise made to him: "Today you will be with me in paradise."[11] He didn't get much time to feed the hungry and clothe the naked, but Jesus put him in with the sheep anyway. This is grace at its most primal. Jesus knows the thief didn't just say the words; his heart changed while he was dying so that he turned into *that kind of person*, the one who would have given up his pants if he had any to give.

Saved by grace is also, ironically, letting go of trying to be good. Trying to be good implies I am in control and can

---

10. John 11:25–26.
11. Luke 23:43.

decide what is good and what is not. It sounds right but it is not: *I'll do good things for Jesus, then He will love me because I am good.*

Grace works the other way around. If I actively believe in the living Christ and seek Him out, He will change me. It is: *Because He loves me, He will make me good.* He will use me to take care of "the least of these" and I won't even notice. He will make me into *that kind of person.* He will herd me over to the sheep.

My friend Mark, another retired physician, volunteers at homeless shelters and free medical clinics two or three days a week. He is kind and generous. He takes care of the home and family needs without complaint when his wife flies away, sometimes for weeks, to serve with Samaritan's Purse. Nearly every week, he volunteers for the childcare program at his church. Here's the kicker, the thing that makes him fit in so well with the sheep Jesus describes: he doesn't do it for Jesus. He does it for "the least of these." I know, because he professes to be an agnostic.

I don't know if Mark is going to be judged with the sheep or the goats. I don't really want him to change anything about the way he lives. But I wish he would profess that He trusts and serves Jesus; then He could also claim salvation by grace. I think it would have a profound effect on his children and their children for generations. And I want to be certain he is herded in with the sheep. I really want him to be one of my buddies in heaven.

How many good deeds do you have to do to go to heaven? How many times do you get to pass by "the least of these" before you go to hell? I don't know. Jesus is sitting on the judgment seat right now. I'm praying for His mercy and claiming His grace.

Part Three

I believe in the Holy Spirit,
the holy catholic[1] church,
the communion of saints,
the forgiveness of sins,
the resurrection of the body,
and the life everlasting.

Amen.

1. Or "universal."

Chapter Twelve

"I believe in the Holy Spirit"

I grew up in a Lutheran church where the Holy Spirit was known as "the Holy Ghost." One Sunday when my brother and I were aged about four and five, rather than Sunday School we went to "big church" with the family, probably on Easter or Palm Sunday, a beautiful spring day. I remember nothing of the service except that Bruce and I were very impressed by the Holy Ghost. Our imaginations ran wild with the possibilities of real, live ghosts. In church! Everywhere! Best news ever!

After church the extended family gathered at our home for a holiday dinner. While grown-ups bustled around the house setting the table and preparing food and exchanging news, we sat on the front steps and talked about holy ghosts. Pretty soon, we started seeing them.

I don't know for sure what my brother was seeing, but once he saw a holy ghost, I was going to see one too. As soon as I saw a flicker of sunlight, I said, "There goes another one."

In a moment, Bruce said, "I see one."

I gave it a three second pause: "I see a holy ghost."

Another beat, then Bruce: "There goes another!"

We went on for a while, sometimes seeing two or three at once, having a heckuva good time. It sure beat church, and it sure beat behaving at the dinner table.

Unbeknownst to us, the extended family grown-ups had gathered behind the screens of the open windows and doors, and they were enjoying themselves immensely! Thus began a family story that has, to my embarrassment, refused to die.

Yes, it was cute. Yes, it was funny. But now, ask yourself, why was it funny? Those two little guys were clearly fooling themselves, and each in competition with the other to be a bigger fool. They could not possibly be seeing holy ghosts because one, there's only one Holy Ghost? Two, the Holy Ghost is invisible? Or three, the Holy Ghost isn't real?

Do you believe in ghosts? Supernatural forces of any kind? Spirits of nature? Demons? My family did not. We were of a northern European mindset that believed in hard work, cause and effect, truth and consequences. We didn't even believe in luck. Nobody gambled.

Although we kids kept a more open mind, you can see how we might have been astounded to hear about a ghost in church. I'd like to say that further education and experience in the church made the Holy Spirit more real to me, but that would be inaccurate. Ours was a faith of liturgy and intellect, believing in the forgiveness of sins by grace and in life everlasting. But how the Holy Spirit fit into the Christian life? Who knew?

Sure, we heard of faith healing and even watched Oral Roberts on TV. And laughed. We heard of crazy people speaking in tongues and handling snakes and casting out demons, and we thought them all charlatans, weak-minded folks given to superstition, or even clinically insane. My experience on the psych ward as a medical student did not disabuse me of these attitudes.

Then one day when I was a third-year neurosurgery resident at Yale, my dad called me about a case. He had been a deacon for our large, downtown, conservative Lutheran congregation for several years. Now, an FBI agent, a church member, had suffered a gunshot wound to his cervical spinal cord and was on a respirator at Hennepin County General Hospital. The agent had regained consciousness and sent a message to the church requesting the elders of the church lay hands on him and pray for his healing. He quoted the book of James to the head pastor: "Is anyone among you sick? Let them call the elders of the church to pray over them and anoint them with oil in the name of the Lord. And the prayer offered in faith will make the sick person well; the Lord will raise them up."[1]

Lutherans didn't do that kind of thing. Our prayers for healing usually had the general aim of accepting our suffering in peace rather than any expectation of physical improvement. Healing came from doctors. Comfort came from pastors. But the FBI agent had a persuasive argument. So the head pastor of Central Lutheran Church recruited Dad and two other guys to go to the hospital and pray for healing. It couldn't hurt, and they couldn't say no.

Dad described the experience to me as strange but oddly peaceful. The agent had thanked them, and they all left wondering if they had performed an act of faith or an

1. James 5:14–15.

exercise in futility. Dad wondered, had I heard of this kind of thing before? What are the chances of improvement with this kind of injury?

No, in four years of medical school, a year of internship, and two years of neurosurgery residency, I had not seen anybody lay hands on a patient and pray for healing. I had no experience with anointing oil. As for prognosis, it was hard to tell without specific knowledge of the case, but gunshot wounds to the cervical spinal cord are usually fatal and the survivors are virtually always complete quadriplegics without hope for recovery. But, hey, prayers are always good for comfort anyway, right? What harm could it do? The agent might have some anger issues with God when he didn't get the result he and the elders had prayed for, but that couldn't last. He would forgive God for not healing him, right?

Time passed and I returned to Minnesota for a visit. I remembered the story and asked Dad what had happened to the agent.

Dad looked a bit surprised, remembering something he perhaps should have told me before. "Well," he said, "about six months after his injury I helped officiate at his wedding at Central Lutheran. He walked down the aisle."

Central Lutheran is a cathedral. The aisle is at least fifty yards long and you could only get to it by walking up a few steps. No way a quadriplegic could do this.

Huh, I thought. I must have misunderstood about his injury because that could never happen. Because laying on of hands and anointing with oil and prayers for healing couldn't do this. Because, much as I believed God was real and Jesus forgave sins, I didn't believe in the Holy Spirit. I didn't believe that God would interfere in the affairs of men beyond changing hearts and minds. He might have created the world, but He didn't heal spinal cord injuries.

The tables had turned since I sat on the front steps and counted holy ghosts while my dad watched from inside the house and laughed because he didn't see any. Now he had experienced the presence and the power of the Holy Spirit while I watched and, although I didn't laugh exactly, I couldn't see what he saw.

That was a long time ago. Since then I've encountered the Holy Spirit at the bedside when people who should have died recovered and people who should have been paralyzed got up and walked. I've encountered the Holy Spirit in the streets where men who were bent on getting drunk heard the gospel presentation, left their booze, and prayed for eternal life. I've encountered the Holy Spirit in the jungles of Central America where I witnessed what can only be described as a demon being cast out of a man.

I think it is possible to think of yourself as a Christian if you believe that God created the world, and Jesus was the Son of God who entered human history and died for the sins of the world and was raised again from the dead. That's the first two parts of the Apostles' Creed. Those beliefs alone bring you the greatest story ever told and lead you into the most gratifying philosophy of life—the one preached by Jesus. That's what I once did.

But when I left it there, I only embraced a story and a philosophy. I did not experience the power and the beauty and the joy of God working actively in the world, in my life, in the right here and right now.

How did I move from the cynicism I had as a young doctor to that of an old guy who believes in miracles? It's not a single story; it's multiple experiences. But if I could point to a beginning, I would point to the first time I heard God's direction clearly in answer to a prayer. "No, you can't fix it," He said. "Just go home."

I related the story before, and I won't repeat it. But that experience opened up one new vision for the universe and one new insight into the experience that follows the forgiveness of sin.

The new vision of the universe is simply the recognition that the material world—enormous and complicated as it is with all the matter, time, space, and energy—is not the complete reality. Perhaps I recognized it before in some abstract way, but I was convinced that that spiritual world—heaven and hell, angels and demons, miracles and curses—was separate from our reality. Then God spoke and the veil between the two worlds broke.

Then comes the forgiveness of sin and what happens next. Think of sin as an addiction to a behavior contrary to a life of love and peace. Think of alcohol and drug addiction. Think of gambling and sex addiction. Then think of the twin addictions of greed and vanity that our culture encourages.

Every addict knows you can't fix your addiction yourself. Over the past hundred years or so, our culture has gone from viewing addiction as moral turpitude to viewing it as disease. Is this a step forward? It is the recognition that the individual is helpless, unable to save himself or herself. It allows for the cultural equivalent of the forgiveness of sins. It is permission to leave the past behind. This is the first step in the twelve-step recovery program modeled by Alcoholics Anonymous.

Although the first step is necessary to leave the past behind, is not enough to provide future freedom from addiction. For that, steps two and three are necessary:

First, recognize that a greater power can restore us to sanity, and second, we can decide to turn our life over to that power.

Welcome to the Holy Spirit. I am a sinner incapable of saving myself. I can study the rules, the moral codes, the ethics statements, the Ten Commandments. But what I know and want to control does not rule my behavior. I am not alone in this. St. Paul says, "I do not understand what I do. For what I want to do I do not do, but what I hate I do."[2] Those of you who have never been trapped in sin or addiction think that statement nonsense; those of us who have been there know exactly what Paul is talking about. To move forward is to turn our lives over to that greater power, and that greater power is the Holy Spirit.

Please recognize that this is more than deciding to become a better person. I tried; it didn't work. This is giving up trying to be good. This is surrendering. It is seeking relationship. It is trusting in something intangible, an otherworldly spiritual power. Some would call it "magical thinking." I call it the Holy Spirit.

This is the essential power of the Holy Spirit, the transformation of sinners into something closer to saints. I'm not at sainthood, but I'm a lot closer than I was when I tried to be good.

There is more, of course. The Holy Spirit has the power to change lives in this world and the next for confessing sinners and those still lost and seeking. He has the power to heal the sick and bring about countless miracles. He is the living spirit of Jesus in the world today. There are specific gifts of the Spirit for each believer. There are no limits.

---

2. Romans 7:15

But there is a guarantee. If I can let the Spirit work in me, I can be the kind of person that shows the "fruit of the Spirit": "love, joy, peace, forbearance, kindness, goodness, faithfulness, gentleness, and self-control."[3]

Now I'm an old man sitting on the front steps of my life, still seeing things. My brother is already in heaven and maybe he's seeing more clearly what I see. I see the Holy Ghost, and there He is again ... and again. I'm seeing an FBI agent walking down his wedding aisle and a brain-dead woman coming back to life and a redeemed alcoholic. I'm seeing me as a jerk in my twenties becoming something better in my seventies. I'm seeing us then and I'm seeing us now and the Holy Ghost has worked His magic. The FBI agent and the woman and the alcoholic and me. We all look a lot more like Jesus.

---

3. Galatians 5:22–23.

Chapter Thirteen

"The holy catholic church"

My first mission trip was to the Philippines with an evangelism team, and I carried my first-aid kit outfitted for international travel. I expected to provide support to the team and to participate only marginally because I was way out of my comfort zone. We traveled for over twenty-four hours and crossed a dozen time zones and the international dateline. When we arrived, I not only didn't know what time it was, I didn't know what day it was. We slept for a few hours, awoke groggy, brushed our teeth, and put on our best clothes because it was Sunday morning in the Philippines. Even though it would be a dozen hours before Sunday morning arrived back home, we were going to church with our host.

Cement floor, cinder block walls ten feet high under a tin roof twenty feet above that, humid air circulating through steel rafters, barely moved by distant fans—this was church in Calinan, outside of Davao City, Mindanao, the Philippines. Six hundred men and women crowded together, singing and dancing to amplified gospel music sung in Cebuano, Tagalog, and English. I felt disoriented and alienated, and then, surprisingly, like I was at home. Though on the opposite side of the globe, I worshipped the same God, the same Christ, with the somehow connected body of believers, the *catholic* church, the church universal. Magic.

After returning home I shared my experience with the OR team while we waited for the patient to arrive. The circulating nurse, Donna, asked me a tough question. I had known her for several years, first as an ICU nurse and then as an OR nurse. From previous conversations I knew she was raised Catholic but no longer attended church.

"Dr. Lohse," she said, "why is the church filled with hypocrites?"

Of course I wanted to argue that this wasn't so. But I couldn't really. I didn't know her experience; no doubt she had a very clear example in mind. And I could not disagree because I had a few examples myself. Jesus had the same outrage at the religious teachers in His time.

The scrub nurse, the OR tech, the anesthetist all waited for my answer. I'd already confessed that I believed in the universal (catholic) church. Did I believe in the *holy* church?

Jesus, on the night before His trial and execution, had the famous Last Supper with his disciples. The last thing He did before they left the room was share a cup of wine with them, and the last thing He said was, "This is my blood of the covenant, which is poured out for many.... Truly I tell you, I will not drink again from the fruit of the vine until that day when I drink it new in the kingdom of God."[1]

I am told that this phrase, so familiar to us in our Communion service, is a paraphrase of the vow given at the Jewish *ketubah*, the finalization of betrothal before a wedding. The bride price and dowry are agreed between the families and the groom drinks a cup of wine and swears to abstain until he is ready to take his bride.[2] Jesus is preparing His followers, His *church*, to become His bride.

You would think that Jesus, being God and all, could have found a better bride.

We create scandals. Congregations fire good pastors who preach a gospel that offends the status quo. Good pastors quit in frustration or depression to find work somewhere else, anywhere else. Boring preachers preach congregations to sleep and to extinction. Congregants have affairs with pastors. Priests prey upon children. Evangelists skim extraordinary profits from fervent converts. Healing ministries offer unrealistic hopes to the uncurable. Spiritual gatherings deteriorate into cults that separate people from their families and exploit them for the slavish adulation of an all-too-earthly authority. Yes, there are hypocrites.

And yet...

We are the holy catholic church.

We are the bride of Christ.

We are the hope of the world.

I grew up in a church, my wife and I were married by

1. Mark 14:24–25. See also Matthew 26:28–29.
2. Jeffrey D. Hoy, "Betrothal," Words of Faith Daily Devotional, March 26, 2024, https://www.faithfellowshipweb.com/blog/post/betrothal.

a pastor of that same church, and our first son was baptized there as an infant. I feel some institutional allegiance to the place, kind of like claiming alumni status from my high school and college. But it wasn't family, it wasn't community; it was an institution. Even after I believed in the ultimate reality of God and the resurrection of Jesus, I believed church membership to be a quaint tradition, like joining a Christian club—a totally optional activity.

Therefore, my church attendance was sporadic, practiced only at my convenience with minimal contribution to congregation needs, and I established only the briefest of relationships with other believers. Outside the church, I worked hard to gain worldly success while I drifted into spiritual darkness. Still, I didn't associate my spiritual state with my church involvement or lack thereof. Eventually, by the grace of God and the love of a good wife, we made it through the darkest of times and ended up in a new city with new opportunities. Mary and I felt a profound sense of gratitude, and we committed to join a church.

Then I sat through many a church service with the well-scrubbed and well-dressed and well-respected. I harbored a secret cynicism that this was a group of good people gathered to look good, and looking good was more important than being good.

The more I hung around, though, the more I saw kind people. They may have been equally well-scrubbed, but when the opportunity arose, they didn't shy away from the down and dirty. They may have been well-dressed, but they went out of their way to do clothing drives and provide underwear and socks to people who didn't have homes. They may have been well-respected, but they didn't ignore the lonely and the lost. On Sundays, quiet heroes worshipped side by side.

I found out I was a baby Christian. I had read only two books of the Bible completely, I had no prayer life, and I had no Christian friends other than my wife. Through this church I slowly learned a basic truth about Christian life: To become more Christlike, you must hang around with other people who are trying to become more Christlike. Otherwise, you drift. You might, like me, drift into thinking that the traditional rules are outdated; that Jesus may have only "spiritually," not physically, risen from the grave; that miracles don't occur anymore; that prayers are futile; and finally, and fatally, believe that you get to personally decide what is best for yourself without the wisdom of ancient biblical guidelines. The drift took me right back to the garden of Eden, eating an apple with a snake.

Peace, joy, and purpose come not from a solitary journey but from a community of love, and the church at its core is a community of love: God's love for His people, and His people's love for each other. If I had one bit of advice to give to anyone, it would be to spend an hour in church each week *even if you don't believe.* If you spend time with people seeking the joy of the transcendent universe and learning how to become better and more loving people, you can't go wrong. If your goal is inner peace and joy and finding purpose, I'm pretty sure it's at least as good as your yoga class, your meditation lesson, or your self-help podcast.

Of course, being a human institution, every church also drifts. Even in the early days, the church drifted from the gospel of grace toward the rules of religion. In medieval times the church drifted from the majesty of

God and the power of the Holy Spirit toward the majesty of Rome and the power of wealth and weapons. In the modern times? I don't really know. We drift all sorts of ways. All I know is sometimes I want to weep on Sunday mornings.

Do we exist as a church to maintain a building and membership rolls? Or do we exist to bring good news to the poor, comfort to the sick, and to set the prisoners of alcohol and drug addiction free? I cringe when I hear those who are hell-bent on saving the unborn and totally indifferent to the plight of the pregnant, or for that matter, those who stridently claim reproductive rights at the cost of infant deaths. We thump our Bibles for law and order, and we thump them again for social injustice. We thump them better than we read them. We are incredibly imperfect. And we are filled with hypocrisy.

One day I sat in church and looked around. I'd been in this congregation long enough that I knew many of the testimonies of the people in the pews. Appearances varied widely from the aging hippie to the aging preppie, from the young and tattooed to the old and frilly. And yes, they were for the most part kind, well-intentioned, sometimes wise, sometimes heroic. But what they had in common was more than their goodness. It was their redemption. I saw alcoholics and ex-cons and serial arsonists. I saw the divorced and the adulterous and the pornography addicts. I saw the greedy and the vain and the selfish. And they were all there raising their arms and singing praise, thanking God for forgiveness and grace, and praying for the strength to be better people, to be more like Jesus.

This is the bride of Christ. This is the hope of the world. This is the church of hypocrites and sinners who are forgiven and redeemed and transformed by the Holy Spirit. And they transform the world.

Anybody who knew me in my worst moments and saw me in that congregation would think, *Yes, there he is, the hypocrite, with all his fellow hypocrites. What's he trying to prove?*

So, yes, Donna, the church is filled with hypocrites. Like me—an imperfect man saved by the grace of God. We are seeking peace, experiencing joy, and learning to love. We aren't there yet. But we are trying to become better; we are trying to become more like Jesus. He keeps embracing us, loving us, putting us back on course. The bride of Christ isn't perfect, but the groom of the Church is. He will make us holy.

Chapter Fourteen

"The communion of saints"

My first Communion terrified me. At age fourteen I "confirmed" my baptism vows, meaning I relieved my parents and baptism sponsors of their responsibility for my eternal soul. Now heaven or hell was completely in my hands. The next step was Holy Communion.

A line of us sat in the first row of the cathedral-like church, the boys in suits and ties, the girls in modest dresses, reciting the liturgy that forgave our sins. It was now time to march forward to the rail and receive Communion. But what if we had thought something impure before we got the bread and wine? What if we had doubts? What would happen to our eternal souls if we still went through with Communion? We whispered among ourselves and wondered. We thought about escaping, but the exit routes seemed to be blocked by stern-looking ushers. Even if we made it out of the church, we would still have to face our parents. Nope. It was a done deal. We had to march up and take our medicine—like all sacraments, for better or worse.

We all survived. We all became saints with a small *s*.

It's a radical thing to take into yourself the body and blood of Jesus. Maybe the bread and wine transubstantiate, maybe the bread and wine are symbolic, or maybe we simply don't have words for a reality that we can't understand. What is clear is that Jesus wants to be part of me. No, that's not exactly right. Jesus *insists* on being part of me. He demanded something unheard of back in His time on Earth. He said, "unless you eat the flesh of the Son of Man and drink his blood, you have no life in you. Whoever eats my flesh and drinks my blood has eternal life, and I will raise them up at the last day."[1]

Shocking. Discipleship ranks thinned considerably.[2] Only twelve guys hung around, and one of them would betray him. "Ease up," I want to say. "It's only bread and wine." But that's not what Jesus said, even when He offered bread and wine. He said, "Take and eat; this is my body." And a moment later, He gave them a cup of wine and said, "Drink from it, all of you. This is my blood of the covenant."[3]

That's Holy Communion, that's the Eucharist, and it is a powerful part of being a Christian. No other religion has anything like it. It is not in the Jewish tradition. In a way, it is repulsive, suggestive of cannibalism—a nearly universal taboo. Only Jesus insists on being literally part of His followers, body *and* soul. This is how we become saints. Jesus is part of me. Jesus is part of you.

Each month our church celebrates Holy Communion, and as part of that service we usually recite the Apostles'

1. John 6:53–54.
2. See John 6:60–66.
3. Matthew 26:26, 28.

Creed. At the point we say the "communion of saints," I get a foggy vision of the ancient church fathers in beards and robes sharing a loaf and a cup. St. Francis or St. Patrick or St. Augustine? I don't know; I'm not good with saint identification. But now I've lived long enough to understand where I'm standing and with whom I'm standing.

I see a guy who used to be drunk all the time and burn down buildings for the insurance money, but now he visits prisons and tells convicts they are loved by their Creator and that they have purpose and worth. I see another guy who used to be addicted to pornography sites, but now he is the father of two daughters and leads a men's fellowship group. I see another guy who used to be a cocaine addict singing in the choir. I see a recovering alcoholic who is finishing his stint as the chairman of the church council. And there's the serial adulterer with the miraculously healed marriage.

And me? I can't even think about me. Not enough pages exist to describe all my failures and all my redemptions. Yet, here I am, standing with the small *s* saints.

Jesus' words that we use in our Eucharist rituals, giving his body as the bread and his blood as the wine, are recorded in the Gospels of Matthew, Mark, and Luke. But in Luke, Jesus shares His feelings: "You've no idea how much I have looked forward to eating this Passover meal with you."[4]

We worship a God who is eager to share a meal with us, the worst of the sinners. He offered it to Judas after Judas had already taken the betrayal money. He insists

4. Luke 22: 15–16 (MSG).

that if we allow Him into our bodies and souls, He will make us saints.

He promises that we'll do it again. He promises a communion of saints now and in the world to come. In the book of Revelation, Jesus says, "Here I am! I stand at the door and knock. If anyone hears my voice and opens the door, I will come in and eat with that person, and they with me."[5]

Once, as a teenager, I knelt down in a cathedral, a suit coat pulling against my elbows and shoulders, a starched collar tight across my throat, hands cupped to receive a Communion wafer and a thimble of grape juice, terrified of hell and damnation, hoping against hope that nothing would change. Now, I hold out my hands, cupped in the same way, surrounded by my friends, the sinners who will soon become saints, confident that what I will feel is the love that heals the world, and hoping against hope that everything *will* change.

---

5. Revelation 3:20.

Chapter Fifteen

"The Forgiveness of Sins"

I was consulted to see a patient in the ICU who was an inmate transferred from the Department of Corrections for a complex illness. I found that he was so bound by shackles and restraints that I could not adequately examine him. I insisted the officers remove some of the restraints during part of my examination. Then I wrote up my recommendations for further tests and arranged for a follow-up visit.

When I returned the next day, the corrections officer's supervisor intercepted me. "Doc," he said, "there's something you gotta know about Prisoner X."

"Okay."

"He killed three people to get into prison and got a life sentence. He wanted a death sentence. So he killed two more people in prison, other inmates. Now he's on death row. He tells us he wants to kill more. So when you ask those guards to take off those restraints, it's a risk—a risk to the officers and a risk to you. If he gets a chance, he'll rip your throat out with his teeth."

Well. I'm sure he meant to add, "And have a nice day."

How do we explain evil? Sometimes, like in the case of the psychopathic multiple murderer, I come face-to-face with it. Sometimes I hear about it in the stories of friends and relatives. Inexplicable evil confronts me every day on any news feed. This end of human history tells the story of man's violence against man. We like to talk about misunderstandings and mental illnesses and cultural differences. We even talk about hate and prejudice and sexual violence. We propose solutions like a war on poverty and better education and more laws against racial discrimination and wider access to mental health care. But the root problem is simply evil. Our world is full of sin and sinners. Ask any district attorney. Ask any beat cop.

In the book of Genesis, sin originated from consuming forbidden fruit in the garden of Eden. I tend to trivialize this and call the fruit an apple and treat it like a dietary no-no, like having a bowl of ice cream before bed: No, you shouldn't do it, it's a bad health choice; the only reason for eating ice cream before bed is that it tastes good at the time and the bad consequences are delayed. There are some good parallels here, but in the Garden, the fruit was not an apple or ice cream but the fruit from the tree of Knowledge of Good and Evil. Eve didn't get ice cream; she got the knowledge of good and evil and all the consequences of that knowledge. This is the origin of sin.

This confuses me. Knowledge is good, yes? Knowing good from evil is very good, yes? I take pride in knowing right from wrong. I teach my children right from wrong. I expect everybody else over the age of five to know right from wrong. Eve ate this fruit that opened her eyes to the knowledge of good and evil. It was so good she shared it with Adam. And what happened then?

They started making fig-leaf clothing, not because they became naked but because they were *ashamed* of being naked. They were alienated from God, not because God didn't visit the Garden, but because they were *ashamed* and hid from God.[1]

Our world is locked into a destructive, moralistic cycle that began with taking into ourselves the fruit of the knowledge of good and evil, an attitude of pride and arrogance that allows us to believe that we make better decisions than God about what gives us joy and purpose. We become ashamed. We are naked, physically and spiritually, alienated from our Creator. And out of this shame and alienation, we commit sins—all the many acts that separate ourselves from our Creator and His creation and all the many acts that separate ourselves from each other. Some acts are little things like eating ice cream before bed, self-indulgences that mostly lead to self-harm. Some acts are bigger, like murdering a few people and planning to rip the throat out of the next person that gets close.

So, yes, I believe in sin. But can I believe in the *forgiveness* of sins?

God sent Adam and Eve out of the garden of Eden. There was no place in paradise for a couple who had that sort of knowledge, but He didn't send them out naked. Nor did He send them out with the pitiful fig leaves that they had made for themselves. God personally made each of them clothes of wool and leather.[2] This is a sign of how He really feels about them. This is the sign of His uncon-

1. See Genesis 3:4–10.
2. See Genesis 3:21.

ditional love. This is the beginning of the story of God's redemption of the world. As soon as we discovered how to sin, God began the process of forgiveness of sin.

Forgiveness doesn't seem like a good plan. What good would it do if the rapist said, "I'm sorry," got his bit of forgiveness, and went out and raped another girl the next night? How about the robber who holds up the convenience store, spends the money on dope, gets high, asks for forgiveness, and holds up another convenience store the next day? Or the megalomaniac who starts a war, commits genocide, steals an entire country's wealth, then goes to confession, says he's sorry, gets the forgiveness chip, and starts planning the next invasion? Forgiveness alone does not stop the dark and dismal cycle of evil.

What does it take for the rapist, the thief, the drug addict, the megalomaniac, the murderer to leave the dark path, turn around, and look toward the light? The dark path is filled with inertia, the feeling that what you've done is what defines you, and that you'll do it all again because that's who you are. Maybe you've already rationalized that what you've done is only reflective of the world around you, and you feel nothing. No remorse. Maybe you are not even lonely because you are now locked into a solid companionship with the rest of the damned.

Is there no hope for them? Are better rules and better enforcement and worse punishment the best we can hope for because we are full of the knowledge of good and evil? That makes sense to me.

At least it makes sense until I consider my own sins, my own regrets. Then I hope for more than my knowledge of good and evil.

I have a closet full of regrets. They are sins committed

without the possibility of atonement. Harm was done. Adultery. Greed. People died or became disabled because I made the wrong decisions for the wrong reasons. *Sorry* is not a big enough word, and for most of the regrets there is no one who could hear it.

Here is one story.

In New Haven, Connecticut, in the late '70's, a big man was a kingpin for the local drug trade and petty thievery. He stood a bit north of six feet and went close to four hundred pounds. One summer night he stood in the middle of Howard Avenue drinking beer and insulting his girlfriend. She took his car, a big Chrysler, drove to the end of the block, turned around, pointed the car down the middle of the street and screamed that she was going to kill him. He laughed.

She revved the engine and raced toward him. He stood his ground and swigged his beer. At the last minute, she swerved and missed him, skidding to a stop at the next intersection. She turned the car toward him again, and she screamed her threats again. He shouted back insults and called her a coward. She revved the engine again and this time she did not stop. She hit him and kept going, driving through the block and the next intersection, his body being dragged under the car before falling off directly in front of the Yale-New Haven Hospital Emergency Room. She never stopped until the police pulled her over.

He was alive, conscious, still loudly profane, and paralyzed from the waist down. Too big to be lifted onto a gurney, an ambulance had to come with special stretchers to get him the last hundred yards into the ER. X-rays barely penetrated his mammoth body, and the CT scan table was not able to take his weight. He had a fracture of his lower thoracic and upper lumbar vertebrae, and his neck X-rays penetrated clearly only down to C3. An ortho-neuro team operated to fix the lower spine, and he awoke now paralyzed from the neck down and dependent on a respirator to breathe. A C5 fracture had been missed. Now he lay in a rotating bed in the Neuro ICU, his neck in traction, intubated on a respirator, and facing a very uncertain future. His girlfriend was in custody on assault and battery charges.

Up until this point, I had been aware of the case but uninvolved. My responsibilities were primarily in the research labs where I was doing experiments, ironically, on spinal cord injury. But Saturday morning arrived, and I also had the responsibility of being the single neurosurgery resident on call until Monday morning. I was formally presented with the case. The attending physician for the weekend, my boss, told me that the patient needed a tracheostomy and that I should do it that day.

A tracheostomy is usually a simple operation, often performed at the bedside with a small procedure tray. Although it is a common ENT procedure, neurosurgeons up until the late twentieth century performed the operation on their own patients when necessary. The order from the attending was not unusual.

Except, I was the only one there for the rest of the weekend. If I spent an hour doing a tracheostomy, no one else would be available for the sixty or so inpatient neurosurgery patients and whoever came in through the ER.

Except, the man did not have a normal neck. His neck circumference was equal to a small adult's waist size. The trachea was not immediately below the skin surface. A tracheostomy would be a challenge for an experienced team in a well-lit operating room.

Except, it was not an emergency. The patient had a functional airway through an endotracheal tube, and if something went wrong with a procedure done over the weekend, there would be no back up.

I told my boss it would be a bad idea. I suggested an ENT consult. He growled that neurosurgeons always did their own trachs, like it was a point of manly pride. I told him I couldn't do it alone. He told me he'd come in to help as soon as I finished my other work.

I finished my other work around 6 p.m., then called him hoping the late hour would dissuade him from spending the evening in the hospital, suggesting again that this was something that could be done on Monday morning by a full OR crew with back up. He showed up at 8 p.m., insistent.

The procedure was the nightmare I had imagined. The lighting was poor, the anatomy was difficult due to the thick neck, the trachea hard to find and control, and none of the standard tracheostomy tubes was large enough to easily fit into the airway and come out to the surface of the skin. Finally, after two hours of struggle we managed to get an airway into place. I secured it with four sutures and a tie around his neck. My attending went home. I cursed him under my breath.

At 1 a.m., the patient coughed violently, and the tube displaced from the airway despite the sutures and ties. I ran to the bedside as his oxygen saturations dropped and tried to replace the tube that we had so recently struggled to place. I called for anesthesia to try to reintubate as I worked. They failed. My gloved hand dug deep into his neck to try to find that elusive airway as I listened to the heart monitor speed to tachycardia, his heart pumping more blood to make up for the low oxygen levels. Then the heart rate slowed as the heart itself lost oxygen. Then it stopped. And I heard only the silence of death and regret.

Did I sin? A legalist could argue that a life lost while trying to save a life is not murder, that while the result was unfortunate, the decision to do the procedure was within the bounds of appropriate medical judgment, and that my culpability was limited because I was doing what my more experienced mentor had decided would be best.

But I should have said no. It may have cost me my job, but it wouldn't have cost that man his life. His girlfriend's charges got upped to murder. There was no one to whom I could say, "I'm sorry." There was no opportunity for atonement. Forty years later I still remember the blood on my hands. I still regret that I did not have the courage to say no. That's a sin.

That is one regret. I have more. If I dwell on these things, I am grieved and paralyzed by my grief. If I harden myself to my faults and failures, I am undone in a different way—I become cold, calculating, afraid to move forward, ready to blame another for either his shortcomings or my own.

I have one hope: that the God who made clothes for Adam and Eve as they fled the garden is the same God who will cover my shame. I hope that the same Father who ran down the road to embrace his prodigal son is the same Father that will embrace me.[3] I hope the same Jesus that forgave the guards nailing Him to a cross will forgive me.[4]

More than hope, I need to *believe* this. Otherwise, I

3. Luke 15:11–31.
4. Luke 23:34.

face darkness and despair, which is just another way of saying hell and damnation. I need to believe in the forgiveness of sins. My sins. My regrets. The ones that still wake me up in the middle of the night.

You may have a few regrets, maybe a few sins, maybe more than a few. You do not need to face darkness and despair. You can believe in the forgiveness of sins. God redeems me; He will redeem you. He redeems the world, one sinner at a time.

Chapter Sixteen

"The Resurrection of the Body"

I have so many questions.

Is my resurrected body the saggy, baggy, old body I have now? That might inspire me to learn how to sew fig leaves together. Will I be the skinny teenager I was sixty years ago, or the cute little blond kid in those old black-and-white photos? What about Mom and Dad? Will they have new younger, perfect, resurrected bodies? Will I recognize them? What about Grandma and Grandpa, who I only ever knew as old people? And what about my cousin Terry Joe, who was born with cerebral palsy and struggled to walk and to be understood for decades. What resurrected body does he get? Will he finally dance and sing?

And is the physical resurrection to take place on Planet Earth in this space-time continuum? If all those saved by grace throughout all generations are resurrected at the same time, won't the planet get a bit crowded? Especially since we will all be living forever?

I would find it easier to believe in a "spiritual" resurrection in which my soul rather than my body becomes alive in a utopian place and I live in communion with the souls of all the other saints. But the Judeo-Christian tradition has always been deeply connected to our physical beings and our literal Earth. Jesus taught us to pray, "Your will be done, *on earth* as it is in heaven."[1] We call Jesus the *Son of God*. But He called Himself the *Son of Man*.[2] The language of the Bible seems rooted in the physical rather than the metaphysical universe.

When I started college at age eighteen, I needed to work. The university had a job service, and this listed the position of *diener* at the University Hospital morgue. I was not yet a pre-med student, so the job had no professional attraction, but the pay was decent and included on-call pay. Surely the reason the job was open was that other students were squeamish about dead bodies, but actual contact with the dead would be minimal. I envisioned washing the counters and sweeping the floors. I took the card for the job referral and one other card for selling automobile tires at a store on Lake Street. I called and the hospital administrator invited me to come for an interview that day. Excellent.

His cubicle was on the third floor of the hospital, the same floor that housed the labs, X-ray department, and ER. He told me my duties would be to assist the pathologist in autopsies, clean up afterward, contact the mortician, and deliver the body. He asked if I would still be interested.

A smart boy would have politely declined, but I have a poker face that masks fear and, at that time, a great reluctance to look like a wimp. "Sure, no problem," came

1. Matthew 6:10. Emphasis added.
2. See Matthew 25:31; John 6:53.

out of my mouth while my brain was calculating whether I still had time to make it to the Lake Street tire store before the end of the day. No doubt the hospital would take my application into consideration and possibly schedule me for a training session the following week.

"Great," the administrator said. "We'll get you started right away." He made a quick phone call to the morgue to tell them we were coming and said, "Follow me."

Again, a smart boy would have...

But my poker face and I followed the administrator to the most distant corner of the lowest floor of the hospital. He knocked on a large wooden door, turned to me and said, "Wait here," then turned again quickly and walked fifty feet down the hall. There he stopped, turned back briefly, and said, "I'm sorry, I really have to go. This place gives me the creeps." And he was gone.

Wow, I told myself. *I didn't know running was an option.* Before I could consider fleeing, the door opened and a diminutive Black man draped in a surgical gown opened the door and said in his Lon Chaney Jr. voice, "My name is Howie. Come in. We've been expecting you."

In the antechamber Howie got me into a surgical gown, and I donned my first pair of surgical gloves. The antechamber opened into two autopsy rooms, both of which were being fully utilized. He took me into the room on the left and explained over the opened body that the pathologist was removing the heart and great vessels. It would help if I would put my hand in the cavity below the throat and pull up so the doctor could see better.

The flesh was surprisingly cold. A few minutes later the pathologist removed the heart and attached vessels that I would later learn was the arch of the aorta. Howie nodded his approval at my performance and then instructed me on how to remove the brain—a procedure I will not describe in detail other than to say that I continued to have a visceral response to the smell of "bone dust" until my last day of retirement forty-some years later. We concluded the day as Howie taught me how to sew up the Y-shaped incision with a large straight surgical needle and a ball of string and clean the residual blood and tissue from the stainless steel autopsy table.

Apparently I passed my final job interview with flying colors. Howie called the administrator immediately after the autopsy and told him I would work out just fine. I am, for better or worse, one of those people who responds to horror and chaos with a cold numbness rather than flight-or-fight panic.

Until that time, I had thought little about death—not real death, up close and personal. I had never touched a dead body. Of course I'd been to family funerals and seen the deceased laid out in their Sunday finest and made up to look alive. I'd read about death and seen a million Westerns and war movies and murder mysteries in which the deaths were noble or justified or tragic, but always neat and clean. Now, I was handling, through thinly gloved hands, organic material that used to be living human beings. Now, I thought about death all the time.

My shifts were nights and weekends. Fortunately, few autopsies are so urgent they must be performed at night, but Saturdays and Sundays could be busy. No pathologist had the time to start their workweek with a body left over from Friday night and a team of clinicians clamoring for answers. Most of the deaths were, of course, old people. These were emotionally easier to take since they were physically so unlike my eighteen-year-old self. They were *other people*. But the young mother who died in childbirth felt clearly tragic, an obvious story of irredeemable loss. My low point was assisting in the autopsy of an infant.

A few months after I started, after a particularly busy weekend filled with cases that were emotionally trying, I had a bad dream. I dreamed my friend M.H. had died. He had been a friend through junior high, high school, and the beginning of college. We were on the swim team together and in the same high school club and college fraternity. He was quiet, smart, athletic, classically good-looking. The dream was simple. I had heard he died, and I kept running into our mutual acquaintances and telling them the news. In each case, they would say, "Oh, that's too bad," and go on with their business as if nothing had happened.

I've had the same feeling many times in real life ever since. A good friend dies, and life goes on although it feels like it shouldn't. It feels like they should be missed more, something should change—a familiar feeling now, but then it was new and disturbing. When I ran into M.H. the next day, it was strange. I almost said, "Oh, I see you're alive." I couldn't tell him about the dream. I gave my notice at the morgue the next day.

Later, during medical school and residency, I witnessed the transition from living to dead, sometimes slowly as a patient deteriorated from their cancer or sepsis, sometimes quickly as the patient bled out or suffered an obstructed airway that could not be opened.

The change is dramatic. The complete loss of any motion, including breathing, the loss of skin color, the open, dulled, sightless eyes—all of it is tragic and alien. Yes, alien. The dead body is so different from the living person it is as if they come from different worlds.

I have no love for the lifeless body. Perhaps because I have a scientist's perspective of physiology and biochemistry and anatomy and a doctor's perspective on death, I can't imagine resurrection. I can't imagine the magic of life returning to a corpse already riddled by injury and disease, cell death and protein degradation already beginning, and the process of complete decay imminent. But maybe that is less of a problem with resurrection than it is with my imagination.

That dead body, so unreal and alien, has just been vacated by something. I never get past the feeling that something beyond the cessation of biological function has happened. Something—inexplicable in the language of physiology and medicine, so real that it is almost tangible—has left the body.

I am not alone in this. We call it the soul, or the life force, or the animus. The better we know the person, the more we feel its absence. Although I care little for the body left behind, I miss the life that has just left. None of my scientific training, but all my intuition, tell me that life cannot just disappear. It lives on in some form. It must be somehow resurrected.

My wife's best friend Kathy suffered from juvenile onset type-1 diabetes. She subsequently experienced every conceivable complication of the disease throughout her entire adult life, including blindness, kidney failure, neuropathy, vascular disease, and coronary artery disease. She also had hearing loss due to antibiotics when a dose was not well-calculated to account for her kidney failure. She underwent many procedures, including years of dialysis, kidney transplants, open heart surgery, and even a heart transplant. Mobility became limited due to neuropathy and vascular disease in her legs. Of all the people I have ever seen, personally and professionally, outside of hospice care, she inhabited the absolutely most diseased physical body.

But her disease did not define her. Jesus did.

I remember Kathy sitting at our kitchen table during one of her visits to Jacksonville. We were both around fifty years old at the time, long enough that she had experienced a quarter-century of visual loss and a long string of other complications, and our family had experienced cancer and other losses. We were no longer naïve about life. Illness had come. Death had passed by, but only for the present; it would return. The religion we had in our twenties had been put through the fire and we were different. She explained something about her faith walk that has stuck with me.

"I realized," she said, "that eternal life starts now."

Jesus said something similar when being pressed for answers about when the kingdom of God would come. He said, "The coming of the kingdom of God is not something that can be observed, nor will people say, 'Here it is' or 'There it is,' because the kingdom of God is in your midst."[3]

Kathy started resurrection life now, and maybe that's the first thing to remember about resurrection of the body. The kingdom is in our midst. Life in the kingdom starts now. *This* is our eternal life. The only change to our "life force" or our "consciousness" or our "soul" after death is that we get new bodies.

St. Paul tried to assure some skeptics, like me, about the resurrection of the body. He used the metaphor of a

---

3. Luke 17:20–21.

seed. "What you sow does not come to life unless it dies,"[4] he said to begin his explanation, and goes on later to say, "as we have borne the image of the earthly man, so shall we bear the image of the heavenly man."[5] This seems plausible and even comforting. This old, scarred body is only a vessel carrying a DNA-like spirit that will be planted into eternity and thrive as something so much bigger and so much better. Still hard to imagine.

Jesus assured us that there would be a life after death. He talked about the kinds of relationships to expect with Him and our Father and each other, but He didn't talk much about the specifics, probably because He knew we didn't have the conceptual framework or intellectual capacity to understand. Only once, when He was being harassed by some folks who did not believe in life after death did He mention some specifics. He said, "Those who are considered worthy of taking part in the age to come and in the resurrection from the dead ... they can no longer die; for they are like the angels. They are God's children, since they are children of the resurrection."[6]

My imagination can handle this. We will be beloved children, and we will be like angels. I can accept that there are questions unanswered, that my humanity and position as a created being limits me from understanding the cosmos in all its complexity, especially that which takes place beyond our space-time continuum and biological life. I'm looking forward to hearing my cousin Terry talk with a clear voice and run with strong legs. I'm

4. See 1 Corinthians 15:37.
5. 1 Corinthians 15:49.
6. Luke 20:35–36.

looking forward to seeing Kathy seeing me. I'm imagining resurrection.

I believe in the resurrection of the body—this one, in this life—and resurrection as an angel-like child of God in the kingdom to come.

Chapter Seventeen

"The life everlasting"

I didn't cry when my dad died. I didn't cry at his funeral, or at his memorial service in Minneapolis, or at his interment ceremony. He was 89, had been sick for a long time, and was much diminished. The euphemism "passed away" seemed quite appropriate. He slipped from this life into the next. I felt like I had spread my mourning out over the previous five years, and I was over it.

About a year after his death I traveled to Minnesota to see other family members. Because the cemetery is near the airport, I decided to stop by on my way home and take a photo of the headstone for my mother. It had not been installed at the time of the burial. I found the plot and snapped the photo. I looked about at the rolling hills and distant trees and passing roadways then unexpectedly burst into uncontrollable sobbing. I could not stop. I could not get in my car and drive away. I could only stand there and weep.

The fact that what had once been Dad was marked by such a small stone overwhelmed me. He led such a big life. Nearly ninety years, from farming with a horse-drawn plow to witnessing space travel, to coming of age in the Great Depression, and serving as an Army officer during World War II. He left behind a sixty-year marriage, three kids, a career, and a legacy of kindness and gentleness. So much, so big, and so gone.

There is no everlasting life in nature. All living things die. Each individual organism, plant or animal, goes through the cycle of birth, growth, reproduction (possibly), and death. No exceptions. We all die.

Yet I have difficulty imagining that death is the end of a life like my dad's. And I am not alone. All major religions conceptualize everlasting life. The Hindu tradition has reincarnation until the unity of the soul with the universe; the Buddhist tradition has nirvana; the Jews, the Muslims, and the Christians have heaven. In the mythologies, various forms of afterlife appear. Despite the certainty of death, all cultures throughout history have imagined everlasting life. A universal longing and a collective intuitive sense tells us that the end of our biological life cannot be the end of our consciousness.

Those of us who have been at the bedside—or the operating room or the street or the battlefield—alongside someone who has died, understand. One moment a person is alive and the next moment, when biological life stops, whether at the end of a ten-year struggle with cancer or a ten-minute resuscitation after a heart attack, the change is dramatic. Yes, the breathing stops and the circulation stops, the skin color pales and the eyelids open over glazed corneas, but the feeling at the moment is more. No matter how much I tell myself it is just the off switch of physiology, it *feels* like something more. It feels like the soul has departed. No wonder that the overwhelming majority of humans believe in life after death.

I believe in death, but I can't help but also believe in life everlasting. I understand death as a biological event in this space-time continuum, but I can't understand life everlasting in the same terms. It is no doubt impossible. As a created one in an immense universe, there is no reason to expect to possess in our limited minds the experience, logical constructs, or language to understand life everlasting. I rely on little glimpses, imagination, and big promises.

My father saw his father in a dream, my daughter-in-law saw her father in a dream, and this is not uncommon. We are convinced we will not be alone.

My friend Jimmy saw his grandmother, a bright light, and Jesus while hovering near death after his gunshot wound. Fellow neurosurgeon, Eben Alexander describes extensive and convincing details about his transcendent experience while appearing to be brain-dead.[1] Many others have related their own near-death experiences. The details vary, but light, darkness, and the presence of a loved one seem to be common to all.

I imagine all sorts of things. Sometimes I imagine golf in the kingdom. Would I always shoot par? Would everybody always win? Would that even be fun? Will I go fishing? Maybe, if my assignment is to create hell for bad rainbow trout. I wonder if God will send me out on assignment like Clarence in the movie *It's a Wonderful Life*, who was sent back to Earth to keep George Bailey from jumping off the bridge.

---

1. Eben Alexander, *Proof of Heaven: A Neurosurgeon's Journey into the Afterlife* (Simon & Schuster, 2012).

I imagine eternal life, maybe purgatory, will be like the movie *Groundhog's Day*, where the main character lives the same day over and over and over again until he finally gets it right. Maybe I'll get do-overs for all the mistakes of my life, reliving them until I get them right.

I imagine there will be dogs in heaven. Well, good dogs anyway. I expect Sophie the goldendoodle and Fluffy the bichon frisè and Mr. O'Shaughnessy the Irish Setter will be there. On the other hand, Bruno, the doberman that lived down the block where I grew up and bit my dad one day as he walked home from the bus stop, can be somewhere else.

Will I be able to sing in heaven? Inability to carry a tune is a hereditary curse in the Lohse family. It will be great to open my mouth and wail out the tune that's in my head and have it come out sounding that way. Maybe that's the way it will be for my cousin Terry Joe with his speech impediment from cerebral palsy and my friend Allan with his expressive aphasia from a stroke. They could not be understood in this world; they will be eloquent in the next.

I imagine something like the hope of Buddhists and Hindus, the ultimate unity of the soul with the universe, transcendence from the material world and all the associated lust, and ultimate consciousness. This sounds good at first. Ultimate consciousness includes awareness of every action I have taken at every moment in my life, and an eternity to reflect on the consequences—for good or evil. I could celebrate the times I got it right but mourn eternally for the myriad times I got it wrong. And I would face an eternity of grief for the sins I have committed, the people I have failed, the suffering I have caused or failed to relieve. I would be terrified to face this except for the ancient and eternal promises of God: "For God so loved the world that he gave his one and only Son, that whoever believes in him shall not perish but have eternal life."[2]

Everlasting life is a gift of love, conditional only on believing that Jesus Christ is exactly who He said He was and living out a life based on that truth.

A few decades after the crucifixion, Paul wrote a graphic description of the promised everlasting life:

"For the Lord himself will come down from heaven, with a loud command, with the voice of the archangel and with the trumpet call of God, and the dead in Christ will rise first. After that, we who are still alive and are left will be caught up together with them in the clouds to meet the Lord in the air. And so we will be with the Lord forever."[3]

Life everlasting will be like life in the garden of Eden. I will have communion with God. We'll take a walk together once or twice a day. I'll have purpose and some tasks, presumably using the best of what I love to do and what I am most gifted at doing, and no rush to get it done; I'll have lots of time. And I will have the love and fellowship of other human beings. Adam only had Eve, but the communion of saints will continue into life everlasting. There will be many playmates. I will be like an angel. I will be a child. These are the things promised.

After this I can only imagine. I'm hoping to see my folks and my brother again, and maybe my best friend, Mike. I hope Mike is there. He was a good friend to me and a good man, but he didn't spend much time in church, and I don't know what he thought about Jesus and salvation. Does God's mercy extend to a guitar-

---

2. John 3:16.
3. 1 Thessalonians 4:16–17.

playing fishing guide and horticulturist who played hard and laughed easily? I hope so.

Although we are told there is no marriage in heaven,[4] I hope to spend some more time with Mary, maybe going over the parts of our marriage here on earth that we got right and resolving the parts we got wrong. We have laughed a lot here; I expect we'll laugh a lot there.

I wonder about time. Will time itself be changed? There must be music in heaven, but how could we have music without time? What will it feel like to have eternal patience with everything? Sometimes I wonder if I'll get bored, but I believe my ultimate purpose cannot be boredom. None of my skills here are likely to translate well; God won't need a neurosurgeon or a boat builder. But He will have something else for me to do.

I do hope I'll be able to sing.

Amen.

4. Luke 20:35.

Acknowledgments

These chapters were originally presented to Crossroad Methodist Church's men's fellowship group, known as the BIC (Brothers In Christ). I appreciate the wisdom and encouragement of these men who bring a wealth of experience into the room each week. Where else could I ever find more than a dozen men who have failed spectacularly and succeeded well, suffered terribly and been healed greatly, agonized in loneliness and found treasure in love? I entitled this book, "I Believe," but I should have called it "*We* Believe" because there is strength in fellowship.

As always, I could not start a project like this, and certainly never finish, without the encouragement of family. No one could ever replace my wife and best beta reader ever, Mary, who has the right combination of critical skill and skillful encouragement. Thanks also to Brie and Jenn who encouraged me to continue.

Anthony Nayagan, author of "Ancient Wisdom and Modern Leadership," came forward during the later stages of writing and encouraged me with the special message that what was contained in these pages was important for the world to hear. It's the kind of thing that keeps me going.

Editors make authors look good by correcting grammar and style and insisting on clarity. Rebecca Faith went beyond the expected by adding biblical references, suggesting new ones that were unknown to me, and challenging my theology when it got in the way of what I considered good narrative. In other words, she not only made me write clearly, she made me think clearly.

Author biography

Dr. Lohse was born and raised in Minneapolis where he received his BA and MD from the University of Minnesota before he did his post-doctoral training in neurological surgery at Yale University. After practicing for three decades in Jacksonville, Florida, he builds small boats, plays with his grandchildren, and writes.

deanlohsewrites.com

Also by Dean C. Lohse, M.D.

Fiction: *The Surgeon's Sin: Atonement*

An allegory. Neurosurgeon Peter Jenson botches a brain tumor operation. With his patient dying, his wife leaving, and his enemies denouncing his ability to practice, he begins the long journey to redemption by making a promise he shouldn't keep to a woman he can't refuse, and seeking the help he needs from a man he can't trust.

Amazon: http://www.amazon.com/dp/1735255904

Also available on Barnes and Noble.com and as an audiobook on Audible

Inspiration: *Thirteen Miracles: A Neurosurgeon Meets Jesus*

The thirteen healing miracles of Jesus described in the Gospel of Matthew are re-lived in the life of a neurosurgeon. The experiences of a modern healer bring insights to ancient stories, as the miracles, old and new, bring inspiration for seekers, hope for the healers, and comfort for the suffering.

Amazon https://amazon.com/Thirteen-Miracles-Neurosurgeon-Meets-Jesus/dp/1735255920

Also available on Barnes and Noble and as an audiobook on Audible.com

Made in the USA
Monee, IL
04 November 2026

67380034R00075